Climate Change and Original Sin

Under the Sign of Nature: Explorations in Ecocriticism
Serenella Iovino, Kate Rigby, and John Tallmadge, Editors
Michael P. Branch and SueEllen Campbell, Senior Advisory Editors

Climate Change and Original Sin

THE MORAL ECOLOGY OF
JOHN MILTON'S POETRY

Katherine Cox

UNIVERSITY OF VIRGINIA PRESS
CHARLOTTESVILLE AND LONDON

University of Virginia Press
© 2023 by the Rector and Visitors of the University of Virginia
All rights reserved
Printed in the United States of America on acid-free paper

First published 2023

9 8 7 6 5 4 3 2 1

Library of Congress Cataloging-in-Publication Data

Names: Cox, Katherine, author.
Title: Climate change and original sin : the moral ecology of John Milton's poetry / Katherine Cox.
Description: Charlottesville : University of Virginia Press, 2023. | Series: Under the sign of nature: explorations in ecocriticism | Includes bibliographical references and index.
Identifiers: LCCN 2022044600 (print) | LCCN 2022044601 (ebook) | ISBN 9780813949734 (hardcover) | ISBN 9780813949741 (paperback) | ISBN 9780813949758 (ebook)
Subjects: LCSH: Milton, John, 1608–1674—Criticism and interpretation. | Literature and science—England—History—17th century. | Literature and morals—England—History—17th century. | Environmentalism in literature. | Human ecology in literature. | Air in literature.
Classification: LCC PR3592.S3 C69 2023 (print) | LCC PR3592.S3 (ebook) | DDC 821/.4—dc23/eng/20221122
LC record available at https://lccn.loc.gov/2022044600
LC ebook record available at https://lccn.loc.gov/2022044601

Cover art: *The Allegory of Air*, Jan Brueghel the Elder. Oil on copper. (Private collection; photo Johnny Van Haeften Ltd./Bridgeman Images)

To my mother and father

Contents

List of Illustrations   ix
Acknowledgments   xi

Introduction   1

1. "Infant Cries": Meteorological Voices in the *Nativity Ode*   21
2. Early Acoustic Theory and the Aural Soul in *Comus*   46
3. The Power of the Air in Milton's Epic Poetry   78
4. "How Cam'st Thou Speakable of Mute": Satanic Acoustics in *Paradise Lost*   95
5. Milton and the Barometer: Climate Change in Pneumatic Science   119
6. "Throttled at Length in the Air": Environmental Warfare and Climate Regained   144

Epilogue   187

Notes   193
Bibliography   229
Index   245

Illustrations

1. Engraving of automatic hydraulic and Vitruvian organs in Gaspar Schott, *Magia Universalis Naturæ et Artis,* vol. 2 (Herbipoli [Würzburg], 1657–58), 306–7

2. Engraving of hydraulic automatic organs with mechanized birds in Gaspar Schott, *Mechanica Hydraulico-Pneumatica* (Francofurtensi [Frankfurt], 1657), 415, plate xli

3. Detail of demonic musicians from Jacques Callot's *The Temptation of Saint Antony,* 1635

4. Woodcut of serpent with mouthpiece and crook in Marin Mersenne, *Harmonie universelle contenant la théorie et la pratique de la musique* ([Paris], 1636), "des Instrumens" 5:279

5. First page of [Carlo Roberto Dati] *Lettera a Filaleti di Timauro Antiate* (Florence, 1663)

6. Page 21 of [Dati] *Lettera a Filaleti di Timauro Antiate,* containing Evangelista Torricelli's letter to Michelango Ricci on June 11, 1644, with diagram of the so-called Torricellian experiment

7. Intaglio print of barometer and the incumbent atmosphere in Walter Charleton, *Physiologia Epicuro-Gassendo-Charltoniana* (London, 1654), 60

8. Woodcut of balance and lever holding the world, in John Wilkins, *Mathematicall Magick* (London, 1648) 1:81

9. Engraved title page of William Gouge, *The whole-armor of God,* 2nd ed. (London, 1619)

Acknowledgments

Paradise Lost makes the unknowable past seem intensely present—and indisputably ours—all while hewing closely to a deeply felt truth. Looking backwards at the development of this book, I find the view hazier, a faded constellation of memories partially eclipsed by a thousand intervening life experiences. Recollecting, however imperfectly, the book's beginnings—in coffee shops and car rides, seminar rooms, office hours, and library stacks—I am reminded of the ample wisdom, love, and counsel that fed its growth and, more than once, prevented me from abandoning the project. Acknowledgment or thanks, "the slightest, easiest, readiest recompense," as Milton describes them, are the very least that my encouraging and patient supporters deserve.

Writing about Milton's poetry at times feels like trying to carve a new path up a famous mountain peak with more unobstructed views and better access than the conventional route. In venturing out, one looks to more experienced climbers to provide directions, establish footholds, and call regularly from basecamp. Such was my extraordinary luck in choosing my PhD advisor—the kind and brilliant scholar of seventeenth-century literature, John Rumrich. John's inexhaustible support throughout the writing of this book, which began at the University of Texas, Austin, as a dissertation on Milton and the science of sound, is best explained by his unbending commitment to mentoring graduate students and his unaffected intellectual curiosity. The seeds of many ideas in the book were sifted over in fascinating conversations with John, who can delight and edify the listener on the finest points of Milton's poetry and milieu. Striving to meet his high standards for rigorous, honest scholarship and straightforward prose helped to ensure the polish of this volume. It is my sincere pleasure to formally acknowledge and thank John for his substantial contributions to the conceptualization of this book, invaluable edits and feedback, and not least, for his good humor and abiding friendship over the years.

During my undergraduate and graduate studies, I had the good fortune of working with several other exceptional scholars of early modern literature whose mentorship, in many cases, lasted well beyond my time in their classrooms and played a crucial role in nurturing this book into being. I owe a great debt of gratitude to Blair Hoxby who taught me the joys of reading Milton, Shakespeare, Blake, and Keats, and whose impeccable advice has guided my scholarship for the past fifteen years. Without the benefit of his keen sense of the field, the quality of this book would be greatly diminished. Hannah Wojciehowski pushed me to think theoretically, probingly, and across disciplines, and encouraged me to broaden the scope of the study, much to its betterment. I am deeply grateful for her unwavering confidence in my capabilities as a scholar; without her empathy and support, completing this book would have been all the more challenging. I wish to acknowledge Eric S. Mallin for his helpful feedback on early drafts of the manuscript, and the lovely J. K. Barret for her incisive comments and wittiness when I most needed a laugh. It gives me great satisfaction to commemorate my first teacher of Milton in these pages, the late J. Martin Evans of Stanford University, who undoubtably was one of the twentieth-century's best, most affable guides to the poet's life and works. If not for Professor Evans's inspiring teaching and his particular efforts to tutor me, the awe for Milton's poetry I still carry with me would never have been instilled in the first place.

I am indebted to the English Department at the University of Texas at Austin for several fellowships that gave me the necessary time and liberty to thoroughly research my dissertation, which in turn greatly informed this book. I am also grateful for having received a National Endowment for the Humanities Long-Term Fellowship from the Huntington Library in San Marino, California, which provided me with a generous stipend, a serene place to write, as well as daily access to its extensive archives. It was a particular privilege to use the Huntington's vast holdings in the history of science, which enabled me to write the densely sourced fifth chapter of this book on Milton and the barometer. At the library, I often had the pleasure of discussing my manuscript with world-class scholars such as John L. Heilbron, whom I single out here in recompense for his sharp insights on the Galilean contexts of the aforementioned chapter. Parts of the manuscript have benefited from the input of participants in several workshops, to name a few, the 2018–19 Huntington Library Long-Term Fellows Working Group, The Early Modern Colloquium at the Claremont Colleges, and the Milton Seminar at the Newberry Library, which graciously invited me to present my research in 2019. I owe particular thanks to Christopher Kendricks, Lori Anne Ferrell, Seth Lobis, and Stephen M. Fallon for their organization of and incredibly helpful responses to these presentations. The attendees of my talk, "'The sound of blust'ring winds': Diabolical Weather and Acoustics in *Paradise Lost*" at the Eleventh International Milton Symposium at the University of Exeter in 2015, deserve thanks for their engaged responses to the paper which became the basis of the fourth chapter and fulcrum of this volume.

Chapters 3 and 4 contain revised materials from previously published journal articles, respectively: Katherine Cox, "The Power of the Air in Milton's Epic Poetry," *SEL Studies in English Literature 1500–1900* 56, no. 1 (Winter 2016): 149–70; and Katherine Cox, "'How cam'st thou speakable of mute': Satanic Acoustics in *Paradise Lost*," *Milton Studies* 57 (December 2016): 233–60. I am thankful to the copyright holders, Rice University and the former publisher Duquesne University Press, for the permission to republish the articles in revised form. I would also like to thank Angie Hogan and Ellen Satrom, editors at the University of Virginia Press, and Serenella Iovino, Kate Rigby, and John Tallmadge, editors of the Under the Sign of Nature series, for their interest in my manuscript and care in seeing it through to publication. I would like to thank the anonymous readers selected by the UVA Press for carefully and judiciously reviewing the manuscript as well as offering instructive insights and suggestions for improvement. Several others deserve mention for the critical role they played either in supporting my research or preparing me to write an academic book. To this end I would thank Andrew M. Cooper, Taylor West, Yana Skorobogatov, and Christopher Warren. My coffee cup is raised to every barista in the Austin area for keeping me caffeinated and on task.

Before a book is published, it first exists in the author's mind and continues progressively on the page. As far as the world is concerned, however, there is no "book"—it isn't yet real. I am most fortunate as a writer to have a family capable of imagining the reality of a book long before it actually resembles one. For this, I give my love and deepest appreciation to my brilliant trio of sisters, Genevieve, Veronica, and Natalie, and also my parents, Allen and Camille, who are matchless in their energy, wonder, and efforts to help others thrive. My parents accompanied me through the writing process, giving not only moral support, but also sharpening successive drafts of the manuscript with their curious questions and keen sense of storytelling. Thank you, Mom, for everything, especially your undeterred confidence in your daughter. To Dad, thank you for your integrity and your boundless spirit of giving. For the love that sweetly blesses my every day, thank you to my "chef-of-staff" and Thomas, our little angel. Lifting me up until I reached the very end, Adrien, you have made all of this possible with your devotion and capacious sense of what I can achieve.

Climate Change and Original Sin

Introduction

Adam eats the forbidden fruit in book 9 of *Paradise Lost,* and a few moments later, the "[s]ky loured, and muttering Thunder, some sad drops / Wept at completing of the mortal sin / Original."[1] In an epic full of innovation, these verses stand out. They depict the darkening of the sky and the earth's first prospect of a storm. They also contain the epic's lone reference to the doctrine of original sin. The striking juxtaposition of these two extraordinary ideas suggests that deeply entwined with the fate of the climate in *Paradise Lost* are the theological implications of the Fall. The angels' alteration of the heavens and the corresponding decimation of the earth's temperate climate are among the most programmatic and severe penalties imposed in the epic. The phenomenon of climate change, from the initial transformation of the weather to the perpetuation of inclement conditions thereafter, parallels the spectacular loss and unfolding ramifications of original sin.

Living in the world's most air-polluted city during the height of the Little Ice Age, Milton had every reason to lament the condition of the air.[2] As historians have shown, declining temperatures produced exceptionally cold weather for much of the seventeenth century, propagating war, disease, and famine throughout multiple regions of the world.[3] In London, a spike in fossil fuel consumption filled the atmosphere with unhealthy and foreboding black smoke. In addition to inflicting widespread bodily harm, these devastating climate changes were thought to endanger the soul. Largely overlooked by environmental historians, the moral implications of atmospheric corruption were central to Milton's early modern worldview, which assumed that air comingles with each organism's sacred "breath of life."[4] Understanding and improving the moral and physical condition of the climate mattered immensely to the deeply religious poet, not least because air is the primary medium of sound and the shared material of music and poetry. Wary of the effects of atmospheric corruption on the purity of sound and air, much of Milton's poetry portrays the challenge of reasserting the integrity of human reason and art in the shadow of an inhibiting climate curse.

In spite of his own poetic identity, which he "shaped around a pure and immortal self," Milton assumed that original sin irrevocably damaged the earth's meteorology and the acoustics of air.[5] According to his synthesis of religious and scientific thought, all atmospheric phenomena, whether classically or mechanistically construed, are subject to a single primeval curse that transformed the climate into an instrument of moral judgment. In demonstrating that climate corruption is interwoven with the moral plight of humanity in Milton's poetry, this book explores several related questions: What does atmospheric corruption mean to Milton and his contemporaries? What does it look like, and, especially, how does it sound? Can humanity resist environmental forces that corrupt the very ingredients of life? Who or what is responsible for the condition of the air? In examining a prominent seventeenth-century poet's evolving representations of the climate, this book also traces the gradual development of ideas about the atmosphere during the seventeenth century—an intellectual "climate change" driven by experimental activity and heralding an ecologically devastating shift in Western attitudes towards the air.

Advancing a theologically inflected view of Nature on the eve of modernity, Milton's poetry captures a pivotal moment of environmental history. The world was about to undergo, as Michael S. Northcott puts it, a "new separation of nature and culture"—the epistemological division of earth's nonhuman elements from the moral and political activities of human beings.[6] The exclusion from Nature of thought or intention, attributes seen thereafter as proper to humans, and the exoneration of "every still possible God of any responsibility for weather phenomena," dismantled the assumption Milton shared with early modern contemporaries that inclement weather and natural disasters serve a moral purpose—to mortify human sin.[7] Natural calamities, according to the new taxonomy, "are neither punishment nor sign, but part of an order that is, literally, meaningless."[8] From the standpoint of the present ecological crisis, however, the modern assumption that Nature is devoid of moral significance appears dangerously naive. As Bruno Latour argues, moderns willfully ignore and proliferate the intermediations of human society and Nature by artificially separating these realms.[9] Despite centuries of sustaining this separation, we are glimpsing anew their innumerable interconnections in the ecological tragedy engulfing modern collapse. Relentless floods, uncontrollable wildfires, and other calamities intensified by anthropogenic climate change manifest the deep implication of moral agency—day-to-day decision-making by witting individuals—in the so-called natural realm.

Distinguishing too sharply between moral agency and the natural world is what got us here in the first place. Lacking a soul, a will, or the ability to portend, Nature ceases to pose a moral or metaphysical obstacle to environmental exploitation. Examining how the climate was perceived prior to "the disenchantment of the world"—what some would call modernity's "original sin"—reveals a surprising alternative idea of humanity's place in relation to the natural world.[10] The last major poet of the English Renaissance and a witness to the Scientific Revolution, John Milton offers a window onto the human-climate relationship at an early crossroads in its history. Despite the burgeoning influence of the mechanical philosophy, a diversity of animistic beliefs about Nature still flourished during his lifetime. Attentive to the New Science, but adhering to a vitalistic concept of Nature, Milton's poetry illustrates the compatibilities and tensions between a worldview that presupposed the moral significance of the climate and scientific trends that increasingly portrayed matter as passive and inert.[11]

Examining the vexed, though tightly enfolded relationship between human morality and climate change in Milton's poetry, this book seeks to lay bare some of the implicit biases we bring to the study of past cosmologies. The classical historiographical designation of early modern fields of natural inquiry as "proto-sciences," for example, portrays this eclectic sphere of activity, which includes medieval and religious points of view, as not fully evolved.[12] The era's interest in the spiritual or vital properties of air has perhaps led to underestimation of the knowledge that Milton's contemporaries possessed about the climate.[13] Far from hindering sophisticated insight about the atmosphere, however, the seventeenth century's susceptibility to ideas from religion and magic afforded a morally conscientious view of the climate. Bringing to light a premodern ecology of the Fall, *Climate Change and Original Sin: The Moral Ecology of John Milton's Poetry* recovers a religious worldview that regards environmental corruption and human behavior as necessarily connected. Spirit and environment, the entwined fibers of the animistic universe, sustain the *moral ecology* of Milton's poetry: the power of the climate to intervene in the moral sphere and, conversely, of moral choice to influence the environment. Speaking to us from the unexpected quarter of radical Protestantism—a perspective now regularly associated with "climate denialism"—Milton's poetry throws light on the surprising entanglements of environmentalism and moral obligation.

In recent decades, critics have drawn parallels between Milton's portrayal of the Fall and modern ecological concerns.[14] Ken Hiltner and Leah S. Marcus, for example, present Milton's epic as an environmental fable: humanity's first betrayal of Nature permanently ends our rooted and harmonious connection to the earth.[15] They bring much needed attention to the ecological significance of losing Eden but miss what distinguishes Milton's myth from a modern cautionary tale. The environmental problem initiated by original sin is not simply that humans alienated themselves from Nature, but also that satanic powers invaded the world and corrupted its elements, particularly the air. Milton shared with contemporaries the prevalent view that storms and earthquakes are the work of fearsome spiritual agents licensed to inflict misery on humans as penalty of sin. The assumption that Satan rules the air prompted the ringing of ritually prepared church bells during storms and suspicion of witches as the devil's meteorological conspirators.[16] My argument discerns in *Paradise Lost* an ecological fall distinct from, yet coordinated with, the human fall and coincident with Satan's gradual embodiment of and integration into all sorts of atmospheric phenomena, including delusive and portentous meteorological effects. Satan exhibits this meteorological power, for example, when he enters the serpent as a "black mist," and, in *Paradise Regained,* when he raises a storm against the Son of God. The actions of the angels in book 10 of *Paradise Lost* align the malignant effect of Satan's presence in the air with the aims of divine retribution. The "wound" that the earth receives as a result of Eve's sin signifies not only humanity's self-imposed alienation from natural harmony, but also the injurious influence of spiritual evil in the natural world.

This book challenges a divide in Milton scholarship that situates his natural philosophy either in a bygone era or at the vanguard of the new experimentalism. Countering the longstanding critique that Milton's science was stale and bookish, studies such as Karen L. Edwards's *Milton and the Natural World: Science and Poetry in* Paradise Lost and Joanna Picciotto's *Labors of Innocence in Early Modern England,* draw parallels between the interpretive work Milton's poetic environments demand of readers and the new experimentalists' diffident and provisional method of doing seventeenth-century science. My argument stakes out a middle ground. Though Milton's learning was not in the skeptical line of the Royal Society's virtuosi, it was current and broadly inclusive, reflecting new developments in meteorology and acoustics and their intellectual precursors. Explaining the poetry's investments in these two fields of natural philosophy, which are typically relegated to the margins of the so-called Scientific Revolution, clarifies Milton's interest in the New Science. He valued experimentalism not chiefly for its renunciation of religious, classical, or occult views, but because it often contained and built on these older forms of knowledge.

Milton's core belief in the original corruption of the atmosphere helps to account for the impressive sonority of his verse, which sounds to many readers like "organ music," or so the critical commonplace goes. T. S. Eliot's famous critique of Milton's poetry, that it favors sound at the expense of visual and other sensory effects, stoked the twentieth-century controversy over Milton's grand style. But Eliot offers unsatisfactory reasons for why, as he claims, Milton overindulges his "aural imagination," speculating that the poet's eventual blindness exaggerated his innate musical sensibility. Irrespective of these biographical circumstances, Milton's view of the Fall as an ecologically devastating event that permanently damaged the atmosphere explains why the poetry frequently contemplates and tests the quality of Earth's acoustics.

THE MEANING OF CLIMATE

Historicizing the idea of climate prevalent during the seventeenth century, a time of cultural flux and irresolution in the philosophy of Nature, reveals a robust and changeable conception that accommodates a multiplicity of views. Milton's poetry illustrates the wide semantic range of the term. Often the word *climate* or a variant thereof signals the characteristics of a place, as in the "torrid clime" of hell whose fiery composition strikes Satan like a physical blow (1.297). This usage alludes to the Hippocratic tradition of airs, waters, and places, and the classical division of the earth's surface into several climate zones: parallel bands of the globe with distinct weather, seasonal patterns, and cultural traits.[17] Elsewhere in Milton's poetry, however, climate is untethered from geography and embodies a more generic sense of the word. In one instance it refers to a region of the sky, in another, the conditions associated with a period of time.[18] Eve's flowers "never will in other climate grow," not merely because of the distance between Paradise and the nether world, but also because the condition of the climate is "other" than what it was before the Fall (11.274). Sin changes the matter of the climate, rendering what was "incorrupt / Corrupted" (11.57). As penalty, the angels "[b]egan to parch that temperate clime" of Paradise, erasing any physical trace of its former serenity (12.636). This fundamental change—the *othering* of Earth's climate—extends beyond a single region to the entire globe and defines a new epoch in time.

Any representation of climate is also a characterization of air. Not a science unto itself, discourse about the air and its phenomena occurs down through the ages across a disparate range of traditions including meteorology, biblical commentary, humanist texts, and demonology. Reflecting the contours of these heterogeneous fields, the premodern notion of air straddles a metaphysical and physical position in the universe, affecting among other things meteorological phenomena, the operations of the human soul, the interpretation of prophetic signs, and the bodies and biding places of angels and demons. During the seventeenth century, the spiritual dimension of the air lingered both in scholarly circles and the popular imagination, but a wave of scientific inquiry into the mechanistic properties of air began to promote a more physicalized conception of atmosphere. The discoveries came swiftly and rattled the world of philosophy. Galileo claimed that air has weight; Continental thinkers proposed that the consonance of any two musical pitches depends on the rate of coincidence between their vibrations of air; Robert Boyle compared the flexible behavior of air to that of a spring; weather watchers in England, Italy, and France recorded the daily atmospheric pressure with an instrument called the barometer.[19] Each new advance in the seventeenth-century sciences of acoustics, pneumatics, and meteorology brought the West closer to embracing the modern conception of atmosphere, evacuated of harmony, subjectivity, and spirit.

The seventeenth century gave rise to our way of thinking and talking about the air, but the terms we use today have drifted from their original meanings. *Air* and *atmosphere* in modern parlance can be more or less equivalent terms. Yet when the word "atmosphere" was introduced into English in the late 1630s, elemental air was by far the more dominant concept associated with the tenuous substance that enveloped early modern life. The new term suggested a jarringly different physical understanding of the world. An early discussion of the subject occurs in a remarkable passage on wind instruments in Father Marin Mersenne's *Harmonie universelle* (1636). Air, he writes, "can be considered in two ways, that is according to its purity," excluding anything in it that is not elemental, and, the second way, "with the mixture of the vapors and exhalations which are ordinarily contained in the atmosphere or the expanse which receives the vapors."[20] The notion that air is mixed together with vapors similar to those surrounding physical bodies such as the moon and other planets leads Mersenne to question what air is and whether it even exists as an independent and irreducible element: "But no one yet knows whether the nature of the air is different from that of water, and whether it is anything else than rarified water, or if it is composed of all the small bodies which evaporate and arise from all the large bodies of the earth, and particularly from the water and from the ground. In the same way it can be said that the body of each animal exhales a certain quantity of vapors all about itself, which makes its atmosphere."[21] From the idea that animals have personal atmospheres, "can be drawn the reason of several particular qualities of plants and animals which harm the health, as it happens that the rotten grape spoils the others."[22] In these passages, we can see that the term atmosphere (spelled the same way in Mersenne's native French as it is in English) invites a reevaluation of air according to corpuscular or atomistic concepts. Given the destabilizing implications of the new concept of atmosphere, it is perhaps not surprising that Milton along with many contemporaries preferred the simple elemental label "air." I have opted to use the word "atmosphere" on occasion in this study, since it captures a relevant shift in the scientific imagination and Milton seems to have been influenced by several of its associations.[23] Of the three main ways that the word functions in this book—as a synonym for air, a shorthand for the sum of air that Aristotelians located beneath the lunar sphere, and in reference to a blanket of vaporous air surrounding the earth and other celestial bodies—only the third instance reflects a firmly historical usage of the term.

An even greater distance separates the modern, scientific notion of atmosphere, central to our idea of climate, from the early modern idea of air. Comparing the two elucidates a premodern view of climate, which has been almost completely obscured by cultural change. The planet's atmosphere, as defined by modern scientists, is spheroidal in shape, situated around the earth, and has a large, relatively stable mass.[24] The classical notion of air, on the other hand, has no fixed quantity or shape, and though it has a "natural place," is often found elsewhere (for example, trapped underground) or in combination with other elements. It covers the same spatial area as atmosphere, but instead of merely providing the backdrop against which weather *happens,* air is a potentially active entity that intervenes in meteorological, social, and bodily processes. Air brings the natural environment into the body through respiration and sensation. This explains, for example, why Banquo's general impression that "the air is delicate" around Macbeth's castle, which he bases on the presence of birds' nests, is preceded by a more visceral, psychological judgment: "Heaven's breath / smells wooingly here."[25] Air implies movement and mediation, an endless oscillation between the vital functioning of organisms and the wider attributes of the environment.[26]

The versatility and motility of elemental air accommodated a surprising early modern assumption—that climate is a realm of voluble and multitudinous sounds. Concomitantly, sound is frequently portrayed during this period as a figment of air and weather. This partially explains why Mersenne digresses on the physical nature of air in a treatise on music, posing questions seemingly irrelevant to the actual practice of musical instruments such as whether vacuum (*le vuide*) lies beyond the air, whether instruments may sound in it, and if they may sound in the highest, rarest parts of the air.[27] Though he was an mathematician, Mersenne, like Johannes Kepler and other contemporaries, found it entirely plausible and moreover useful to imagine musical sound ringing out in the farthest reaches of the universe. Similarly, Shakespeare's *The Tempest* describes the climate of Prospero's island as full of "sounds and sweet airs" and "a thousand twangling instruments." In countless Renaissance sonnets and metaphysical verse, lovers raise tempests with their sighs.[28] "Wanton winds" utter "mild whispers" in Milton's poetry, and likewise, airs "attune" themselves to the melody of spring (*Paradise Lost*, 4.265).[29] This book proposes that Milton greatly heightens and expands the pervasive early modern identification between sound and air. Throughout his poetry, breezes, thunderclaps, mists, and exhalations embody and overlap with aural and oral effects such as breath, voice, speech, music, and even the noise of artillery.

For Milton and contemporaries, the association between the climate and sound is not an empty poetic or metaphorical relationship. The Renaissance inherited a synthesis of ideas from philosophy and Christian theology in which air suffuses the physical world with vital breath and song. Air is among the elements that the pre-Socratics regarded as the *prima materia* or original stuff of the universe and is akin to the universal pneuma or "breath," which, according to Stoic philosophy, pervades and unifies the living cosmos.[30] Pneuma, which the Church Fathers translated as spiritus, occurs in animals in the shape of the *psuchē* or soul, of which the governing part in human beings is mind.[31] The Renaissance scholar Robert Burton articulates the ancient relationship between *psuchē* and pneuma as a simple correspondence of the body's internal spirits to the condition of the air: "Such as is the air, such be our spirits, and as our spirits, such are our humours."[32] Separately, the ancient Pythagorean tradition of the harmony of the spheres proposes *music* as the central link between macrocosm and microcosm. Because its mathematical ratios reflect the harmonic design of the universe, music penetrates the soul and assimilates man to the whole cosmos.[33] As a common ingredient of both pneuma and music, air is the point of convergence between these two key traditions. By virtue of air's similarity to cosmic breath or spirit, the climate assumes the aural properties of the *musica universalis*. By virtue of the relationship of music to air, the pneumatic soul is particularly affected by music and sound.[34]

Renaissance Neoplatonists developed the notion that air, acting as a powerful vehicle of music and aural sensation, can stimulate and potentially transform the soul.[35] In his *Three Books of Occult Philosophy* (1510; augmented in 1533, translated into English in 1651) Heinrich Cornelius Agrippa, for example, supplements traditional vitalistic and musical explanations with concepts from natural magic to show that contracting influences from the air, for example, speech, can affect the soul:

[Air] is a vitall spirit, passing through all Beings, giving life, and subsistence to all things. . . . Hence it is that the Hebrew Doctors reckon it not amongst the Elements, but count it as a *Medium* or glew, joyning things together, and as the resounding spirit of the worlds instrument. It immediatly receives into it self the influencies of all Celestiall bodies, and then communicates them to the other Elements, as also to all mixt bodies: Also it receives into it self, as if it were a divine Looking-glass, the species of all things, as well naturall, as artificiall, as also of all manner of speeches, and retains them; And carrying them with it, and entering into the bodies of Men, and other Animals, through their pores, makes an Impression upon them, as well when they sleep, as when they be awake, and affords matter for divers strange Dreams and Divinations.[36]

The air retains traces of spoken words and other matter that pass through it, rendering them communicable to whomever the air touches. This and related notions from the domain of natural magic, for example, the idea that song and utterance can harness unseen forces in the natural environment, provide a source of inspiration and concern for Milton who is simultaneously interested in the cosmological powers of poetry and alarmed by the possibility that atmospheric contamination may flow through his work.

Milton balances these compelling psychological arguments from Scholastic, Neoplatonic, and occult traditions, with the emergence during the seventeenth century of new kinds of inquiry on the air. The Aristotelian notion that the atmosphere demarcates a purely speculative zone was gradually replaced in natural philosophy by the idea that air can be measured, manipulated, and tested. One of the first phenomena related to the air that attracted this new empirical approach was sound.[37] Throughout the Middle Ages, the ratios that make up the Pythagorean consonances—musically pleasing intervals such as the octave, the fourth, and the fifth—were seen as the most significant properties of sound. Indeed, since Pythagoras apparently concluded that these numbers were mystically involved in the structure of the universe, the concept of harmony was thought to organize every aspect of life.[38] After 1600, natural philosophy and music became less speculative and more firmly grounded in practice. The ancient concept of consonance and other measures of musical sound were revised in the late Renaissance as philosophers turned increasingly to mathematics, mechanics, and experiment to investigate natural phenomena.[39] As one of the most scientifically literate poets of the seventeenth century, Milton must have been aware of these developments. His statement that in the 1630s he often went to London "to learn something new in mathematics, or in music, which at that time furnished the sources of my amusement," suggests that Milton was not content to know the status quo but sought out the newest discoveries in these fields.[40]

During Milton's lifetime, however, his notion of sound as both a material and spiritual phenomenon of air was increasingly threatened by the mechanical philosophy. A few lines from Descartes' *Principles of Philosophy* (1644) convey the hostility of the mechanical school of thought to Milton's monism: "we perceive by our senses nothing in external objects except their figures, sizes, and movements.... [L]ocal movement not only produces the feeling of titillation or pain, but also that of light and sounds."[41] By this account sound is not an entity that exists independently of the perceiver; it has no distinct material or spiritual properties by which thought or feelings are borne into the mind; sound is merely the brain's translation of a motion of the air that in turn moves certain nerves in the body. Mechanical explanations gradually became the norm for investigators in the new science of acoustics, an informal area of study pioneered by experimentalists familiar to Milton, such as Francis Bacon and Galileo Galilei. Milton draws on these theories to depict the moral confusion inflicted on individuals by the fallen climate. The Fall introduces mechanism and artifice into aural processes, separating sound and meaning from their point of origination. Ardently opposed to mechanical materialism and wary that it undermines the spiritual ontology of poetry, however, Milton balances these views with older traditions to suggest that a purely physicalist view of acoustics fatally discounts the corruptive potential of sound and blunts its Orphic power.

Milton's concerns about the fallen climate go well beyond the potential for corrupted air to interfere with his art. Salvation itself is at stake. As Nathan Johnstone argues, the Protestant concept of the devil "elevated internal temptation into the most important and dangerous aspect of his agency."[42] This concept of the devil is born out, for example, in the temptations of Eve in *Paradise Lost* and Jesus in *Paradise Regained*, some of the most nuanced scenes of diabolism in English poetry. Yet, interpreting these episodes as "internal" temptations leaves out half of the story. Satan's attempts to subvert the interior person, I argue, depend on his immanence *outside* in the physical climate. In Burton's demonology, for example, widely held beliefs about Satan's dominance over the air and ability to manipulate meteorological phenomena cohere with the supposition that he can influence the human mind. Acknowledging that the devil's proper place is commonly associated with "the ayre, all that space betwixt us and the Moone" and affirming that demons "can corrupt the Aire, and cause plagues, sicknesse, stormes, shipwracks, fires, inundations," Burton explains, quoting the Dutch physician Levinus Lemnius, why tempestuous weather renders the soul especially vulnerable to demonic interference: "the devil many times takes his opportunity of such stormes, and when the humours by the Ayre be stirred, he goes in with them, exagitates our spirits, and vexeth our Soules."[43] In discovering new affinities between Milton's thought and the fields of meteorology, acoustics, and demonology, as well as a variety of other early modern traditions, this study reveals the corruptive influence of the environment in key episodes of spiritual affliction and temptation in Milton's poetry. Through corrupt sounds and speeches, portentous meteors such as fogs, clouds, lightning, and squalls, satanic agents of air corrupt and mislead humanity, penalizing mortals with the very climate ills that human sin brought into the world.

THE ORIGINAL CLIMATE CHANGE

The traditional label for the evils brought about by original sin, "Adam's curse,"[44] is much too general to convey Milton's peculiar focus on the climatological aftermath of human transgression. To discuss the nameless earth-transforming event that Milton reconstructs in his poetry, we need words that live up to the idea. Our current environmental crisis affords a perspective from which it is possible to appreciate the moral gravity of shifting climate conditions. For decades we have known that rising CO_2 emissions and other greenhouse gases generated by human activity are driving global warming and changing the earth's climate.[45] The UN's 1994 international environmental treaty officially defines the phenomenon of climate change as a sustained variation in the average state of the climate attributable "directly or indirectly to human activity that alters the composition of the global atmosphere."[46] Whereas this notion would seem thinkable only to a technologically advanced postindustrial society, early modern people anticipated not only the basic concept of climate alteration, but also its human origination. As early as 1536, Martin Luther, for example, avers that the actions of the first humans permanently changed the air as well as everything else in Nature: "For my own part I entertain no doubt that before the sin of the fall the air was more pure and healthful, the water more wholesome and fructifying, and the light of the sun more bright and beautiful."[47] Related to the idea that original sin diminished the glory of Nature is the common "peccatogenic" (sin-generated) view of natural disasters. From this Providentialist perspective seventeenth-century Puritan colonists, for example, often interpreted New England's extreme winter weather as an indictment of their moral failings.[48] Milton's poetry consolidates these prevalent early modern views about storms and divine judgment around a single, primeval disaster—an act of human disobedience that provoked the climate into an interminable state of outrage.

For better or worse, today's climate crisis furnishes the language critics previously lacked to describe Milton's early modern insight. Climate change is a living illustration of the early modern idea that weather and atmosphere embody the moral implications of human actions. In its anthropogenesis, brutality, global scale, and grip on future eons, climate change is a striking apparition of the premodern belief in the ravaging effect of original sin on the earth's atmosphere.

Most of Milton's major poems contend with the problem of the fallen climate. Even those that predate the Restoration masterpieces by several decades, for example, "At a Solemn Music" and the *Nativity Ode,* assume the damaging effect of original sin on the atmosphere—especially on its ability to substantiate exalted music and sound. However, only one of Milton's works directly imagines the climatological fall itself. The following brief analysis of the global and irreversible climate change depicted in *Paradise Lost* will clarify this study's broader critique of atmospheric corruption in a selection of Milton's major poetic works.

The effect of original sin on the earth's climate in *Paradise Lost* is not immediate. Unlike an asteroid striking Earth—the cause of a catastrophic "impact winter" 66 million years ago—the epic's cataclysmic climate change occurs gradually through the staggered actions of multiple, independently willed agents. Satan and his conspirators, Sin and Death, are a primary source of this atmospheric corruption. The description of Satan lurking near solitary Eve in the garden as a "storm so nigh," literally anticipates an impending meteorological disaster and suggests that Satan conducts the temptation according to his scripturally designated role as "prince of the power of the air" (9.433).[49] After Adam and Eve disobey, the atmosphere immediately loses goodwill towards humanity becoming, in a word, inclement. This bad weather grows and persists because of deliberate measures taken by God's angels in book 10 that transform the climate from a blessing to a scourge:

> The sun
> Had first his precept so to move, so shine,
> As might affect the earth with cold and heat
> Scarce tolerable, and from the north to call
> Decrepit winter, from the south to bring
> Solstitial summer's heat.
>
> (10.651–56)

The angels incorporate additional flaws into the regular motions of the universe, including the malign arrangement of the planets and stars, the tilting of the globe (alternatively, the deviation of the path of the sun), and the circulation of disease-bearing and corrupt vapors of the earth.

Despite divine interventions in the final stages of the Fall, the process of climate corruption conjures up the wild, destructive energy of chaos and, at its worst, hell. After the angels reorder the heavens, for example, the winds burst forth and barrel towards each other from their opposing corners of the world (10.691). The bluster falls on the human realm below where their adverse blasts "rend the woods and seas upturn" (10.700). After describing these deadly countermotions, the passage abruptly concludes, "[t]hus began / Outrage from lifeless things" (10.706–7). The personification of outrage—paired with "Discord," who is introduced at the end of the line—appears nowhere in the epic except in this passage. The seizing of the environment by this passionate faculty sheds light on the fundamental relationship between original sin and climate change.

Why does Milton choose the allegory of outrage to portray Nature's new, fallen mentality? According to prevailing definitions, the word *outrage* signified disorderly behavior,[50] a "violent clamor," or "outcry."[51] The passage on the unleashing of the winds is full of words such as "gust," "[b]ursting," "rush," "blast," "loud," and "noise," that convey this violent and disorderly sense of outrage (10.697–99, 701, 704, 705). Milton uses the variant "outrageous" to portray parts of the cosmos that embody a similar dynamic of confusion, rupture, and cacophony. When the Son returns to Heaven after judging Adam and Eve, for example, the narrative shifts to a view of Sin and Death sitting at the infernal gates "that now / Stood open wide, belching outrageous flame" (10.232). Similarly the primordial matter that created the universe is "[o]utrageous as a sea, dark, wasteful, wild" (7.212). To release and exploit the explosive energy of chaos, Satan extracts the ingredients of gunpowder from Heaven's soil and mutilates the air with the "outrageous noise" of cannon fire (6.587–88).

More than merely jolting into chaotic motion, however, the weather is reacting to a crime. *An outrage*, in another sense, is a violent wrong committed against a person or society in general.[52] The word appears in the context of two infamous crimes depicted in the Hebrew Bible. To illustrate the "injury and outrage" associated with Belial's activities in cities, Milton alludes to the biblical stories of Gibeah and Sodom (Genesis 19; Judges 19) in which women are offered up for rape in exchange for the preservation of men (1.497–500, 505). In a much less literal way than suffering rape on behalf of others, an individual or community may be said to *feel* outraged on behalf of a victim. As early as the late sixteenth century the term began to be used in this emotional sense to signal a subject's indignation in response to an injustice or offense.[53] The notion of righteous outrage respecting an affront to oneself or others is useful for organizing the stages of climate transformation into a sequence of physical, emotional, and ultimately moral events. Original sin in its first commission lands like a violent blow on Nature, inflicting a "wound," eliciting "pangs" and groans, and leading to meteorological grimacing and tears (9.782, 1001–3). Sin and Death pile on to the injury, and by the time angels ratify these changes, Nature's reaction has matured from a feeling of anguish to a sense of fierce indignation—call it moral outrage—at the universal death sentence incurred by humankind. Nature blames Adam and Eve, and God is on its side.

In *De Doctrina Christiana*, Milton addresses the problematic disproportion between the singular crime of original sin and the universal curse in two key ways: first, by emphasizing the magnitude of the initial transgression and second, establishing the guilt of successive generations. Thus, he explains that in disobeying God, Adam and Eve perpetrated multiple crimes bundled into one: "what did man not perpetrate . . . each of them a killer of their own progeny (the whole human race), a thief, and a preyer on what was not theirs" (*CW* 8.1:413). Though victims of atrocity, humanity is still guilty in Adam and Eve and deserves to share in the penalty. He points to instances from scripture where individuals commit sacrilege and God destroys their entire offspring: "[s]o in the flood, so in the incineration of the Sodomites, the annihilation of Korah" (*CW* 8.1:415). The implication is that God uses natural disasters—a catastrophic flood, a sulfurous storm, the sudden caving in of the land, a fire—to punish individual sins that rise to a communal or symbolic level. Adam gets a foretaste of this climate-powered justice—we might call it divine environmental justice—after the angels intervene in Book 10. He gazes at the unprecedented deterioration of the climate and exclaims in despair: "O miserable of happy! Is this the end / Of this new glorious world" (10.720). The earth is spared immediate extinction, but a more depressing thought ensues. Adam has traded humanity's glorious future for something far less valuable, even poisonous: a life imbued with "sense of endless woes" (10.754). He questions the divine justice of this hereditary curse:

> Ah, why should all mankind
> For one man's fault thus guiltless be condemned,
> If guiltless?
>
> (10.822–24)

Ours is an age wracked by Adam's question.[54] Despite the fact that a large proportion of emissions are produced by a few wealthy nations, the burden of climate change is shared by the whole world, not excepting comparatively guiltless people and other species. Tragically, too, it affects "all ages to succeed," younger generations who are fighting to have a say in a fate largely determined before they were born (10.732). By portraying his offspring as innocent victims, Adam echoes Milton's first defense of the disproportionate curse. Robbery and murder of whole the human race are the unthinkable consequences of Adam and Eve's actions. He also shares in Milton's second insight from *De Doctrina* that climate crimes belong to everyone. "All that I eat or drink, or shall beget, / Is propagated curse," Adam bewails (10.728–29). Activities as biologically necessary as consumption and generation only add to the problem in an ecologically fallen world.

The weather has always inspired prophecy. Today, climate science generates an untold number of computer models and projections, all geared toward describing and influencing the future. For Adam, too, ominous changes in the skies spark a virtual dialogue with his future sons. These apparitions spew invective at him: "'Ill fare our ancestor impure, / For this we may thank Adam'" (10.735–36). One disobedient child, a version of himself, demands to know, "'Wherefore didst thou beget me? I sought it not'" (10.762). Adam sardonically characterizes death as their inheritance: "fair patrimony / That I must leave ye, sons" (10.818–19). He is able to imagine his unborn sons long before Michael brings them before his eyes because appreciating the meaning of climate change requires foresight. Ripple effects from "changes in the heav'ns, though slow," which invisibly set in motion "[l]ike change on sea and land" may long outlast Adam's lifetime and inflict what Rob Nixon calls "slow violence" on the environment (10.692–93).[55] By contemplating the long shadow that immediate climate changes cast into the future, Adam perceives that death

> [w]ill prove no sudden, but a slow-paced evil,
> A long day's dying to augment our pain,
> And to our seed (O hapless seed!) derived.
>   (10.962–65)

In arguing with his wife, Adam seems to resign himself to their new ignominious title as the first sinners. When he reconciles with Eve, he counsels them not to "blame / Each other," since they are "blamed enough elsewhere" (10.958–59). At first, it is not obvious what he means by "elsewhere." Where does the blame come from? The outrage of the fallen winds points us in the right direction. Recall that Milton describes this swelling of the winds as an outburst "from lifeless things." Yet, prior to this point, winds and exhalations are portrayed as organism-like and perceptive. Raphael insists that air requires nourishment and must be fed, and, in their morning orison, Adam and Eve address the earth's "mists and exhalations" as if they will hear and carry out their prayers (5.414–17, 185). Why, with the introduction of "Outrage," does the text now refer to these vital elements as "lifeless things"? One possibility is that the winds become so furious that their previous state appears lifeless by comparison. However, in Milton's monist universe—as opposed to a Hobbesian one—motion, or lack of motion, are not definitive criteria of life or death. Milton's winds were not dead before the Fall, but outrage has awakened in them a novel form of volatility. Earlier in the epic, when Eve defends her efforts toward self-determination, she says she was once a "lifeless rib" in Adam's side (9.1154). Earth's "lifeless things" include *its* unborn generations, riblike in their inability to stop original sin and environmental corruption. Blame comes to Adam on time-traveling winds from people who live "elsewhere," that is, not in Paradise.

After the Fall, weather is proleptic. The air becomes a medium that can embody future pain and suffering. The abruptly darkening sky in book 11, for example, which accompanies the hunting activities of the lion and the eagle, foreshadows the expulsion of Adam and Eve and all their posterity from the garden:

> nature first gave signs, impressed
> On bird, beast, air, air, suddenly eclipsed
> After short blush of morn.
>
> (11.182–84)

An *impression* is an early modern synonym for *meteor*, a category of natural phenomena that includes comets, weather, metals and minerals, and even earthquakes.[56] These first meteorological signs thus usher in an age in which Nature's prodigies prefigure the affliction etched by original sin on future generations. Human-caused climate change is the reason the weather acquires ominous moral significance. As a result of environmentally abusive behavior, the world is witnessing not only a deterioration of the weather, but also a shift in the way we perceive it. The weather is no longer incidental to human affairs, rather it exists in a direct and apparently hostile relation to them. The latest ferocious wildfire or hurricane is less of a seasonal blip in the climate than an indictment of our way of life and a sign of greater catastrophes on the horizon. And yet we are still eating of the fruit. Scientists warn that the world may imminently surpass climate "tipping points" which could have grave and cascading consequences for Earth systems.⁵⁷ The Fall is the fictional twin of our own era. In both climate myth and science, human wrongdoing intensifies the weather, turning its previously benign motions into dangerous and globally meaningful portents.

This book offers substantial new interpretations of a group of Milton's major works, the *Nativity Ode* (1629), *Comus* (1634), *Paradise Lost* (1667), and *Paradise Regained* (1671), whose dates of publication span several decades of transformative experimental activity in acoustics and meteorology. With the first five chapters focused alternately on the meteorological and acoustical dimensions of Milton's poetry, the book's organization instantiates the interrelation between these two fields as foci of early modern climatology. Of the six chapters summarized below, the first two present readings of the earlier poems. Chapters 3 to 6 provide analysis of the epics.

Chapter 1, "'Infant Cries': Meteorological Voices in the *Nativity Ode*," demonstrates Milton's fundamental wariness of the connection between the human voice and spiritual agency implicit in the atmosphere and explicates his use of classical meteorology to underscore this connection. Making sense of a cosmographical zone that, on the one hand, furnishes the aural materials of poetry and, on the other, carries the communications of false gods, the most celebrated work of poetry from Milton's youth, the *Nativity Ode,* lays out the rudiments of his evolving perspective on the corruption of the atmosphere and the liabilities it poses to the human voice. The chapter initially focuses on Milton's university exercises (disputations and orations on preset subjects) to establish his knowledge of classical meteorology, a tradition rich with oral vocabulary for theorizing natural processes. Aristotelian, Senecan, and Plinian explanations of meteorological phenomena recur in the *Nativity Ode* and help to convey the dual ecological and acoustical stakes of the Incarnation. Long abused by the deceiving voices of the pagan gods and now governed haphazardly by corruptible meteorological motions, the atmosphere's temporary surrender to heaven's music and the becalming Peace it sends down through the clouds portrays the nativity as an anticipation, but not the fulfillment, of a future divine reordering of Earth's unruly atmosphere.

Chapter 2, "Early Acoustic Theory and the Aural Soul in *Comus*," shows that the moral test at the heart of Milton's 1634 Ludlow masque engages with Baconian acoustics and its reevaluation of traditional theories of hearing, sound, speech, and music. Thematically oriented towards the ear, the masque (known as *Comus* after the eponymous villain) garners an entirely new meaning in juxtaposition to this emerging area of natural philosophy. Bacon's acoustical program, outlined in the posthumous *Sylva Sylvarum: or a Natural History in Ten Centuries* (1627), broke with tradition by bringing music into the orbit of experimental philosophy. Though Bacon denounced spiritual magic, his synthesis of music and physics, like many of the "new" seventeenth-century sciences, borrowed heavily from the natural magic tradition. Milton's definition of sound is close to Bacon's but acknowledges that malevolent figures such as Comus may manipulate the airy material of sound by means of sorcery—a possibility Bacon fails to take seriously. Milton's masque calls attention to this blind spot by confronting representatives of the new Baconian acoustics with its magical heritage. In the contest that the masque stages between the exploratory Baconian acoustics employed by the children and the musical enchantments associated with Comus, the former is shown to have limitations, in particular, its defenselessness against spiritually polluted climates and their effect on human sensation.

In *Paradise Lost* and its sequel *Paradise Regained*, Milton provides a compelling mythic explanation for the acoustical and atmospheric corruption depicted in earlier poems. Chapter 3, "The Power of the Air in Milton's Epic Poetry," argues for an expanded interpretation of Milton's portrayal of the Fall that includes irreversible climate corruption. Milton's account of the ecological change brought about by original sin recalls a common notion among premodern Europeans that Satan rules the air. Explaining the history of this tradition, which is based on the scriptural identification of Satan as "the prince of the power of the air" (Eph. 2:2), the chapter demonstrates Milton's use of the doctrine to characterize the meteorological powers of the fallen angels and the atmospheric corruption Satan initiates by invading the world. Two contexts are of particular interest: the Scholastic tradition that invokes the doctrine of the prince of the air to explain the demonic source of storms and an Augustinian school of thought that associates Satan's place on earth with the middle region of the atmosphere. Depicting the fallen angels as weather makers, the commixture of their air-like bodies with Eden's ecological phenomena and the anatomy of human beings, and finally in *Paradise Regained*, the encampment of the devils in the middle air, Milton's epic poetry draws on orthodox beliefs concerning Satan's meteorological agency to explain the origination of inclement weather and its function in the fallen world as a moral judgment against sinful humankind.

Building on the profile of Satan presented in chapter 3 as the embodiment of the power of the air, chapter 4, "'How Cam'st Thou Speakable of Mute': Satanic Acoustics in *Paradise Lost*," explains a mysterious crux: how does Satan cause the serpent in the Garden of Eden to speak? Discovering a striking resemblance between Milton's description of the talking snake and a Renaissance wind instrument called the serpent, the chapter reveals that corrupted sounds are a key environmental manifestation of the Fall. Satan's vocalization of the snake fits into a representational pattern in the epic that depicts Satan as a maker and operator of musical, mechanical, and bodily organs. Satan's meteorological attributes, discussed in chapter 3, play an unexpected role in capacitating this acoustical output. Chapter 4 contextualizes the connection between acoustics and meteorology with a variety of illuminating perspectives from natural philosophy, alchemy, church history, and music history. It also returns to Milton's interest, explored in chapter 2, in the identities and oppositions embodied by the new experimental philosophy and its occult precursors. The figure of the pipe organ in *Paradise Lost,* book 1, for example, a metaphor for the construction of hell's Pandaemonium, recalls the machines in John Wilkins's *Mathematical Magick* (1648)—a document of the New Science that makes explicit its debt to magic. By portraying satanic acoustics as both magically and mechanically engineered, Milton suggests that advancements such as Mersenne's mechanical description of pitched sound must be coupled by belief in the ability of spirits to interfere with sound and hearing.

Chapter 5, "Milton and the Barometer: Climate Change in Pneumatic Science" turns from theological and acoustical contexts of chapters 4 and 5 to a third source of knowledge about the atmosphere and a crucial influence on Milton's representation of original climate corruption. This area of inquiry, referred to by contemporaries as "pneumatics," came to prominence in England with Robert Boyle's experiments using an air pump in the 1650s and '60s. One of the more memorable outcomes of this line of investigation was Boyle's Law of Gases, a cornerstone of modern chemistry. But prior to its codification in scientific theory, pneumatic experimentation centered on the barometer, a mid-seventeenth-century invention that caused a stir in European philosophical circles and revised a dominant Scholastic view about the weightlessness of air in its natural place. Tracing Milton's interest in this scientific activity, the chapter begins with his visit with Galileo, alleged to have taken place near Florence sometime in 1638–39. Just a few years later, the barometer would be invented by Galileo's protégé, Evangelista Torricelli. The chapter contends that Milton's contact with Galileo and close friendship with Carlo Dati, another student of Galileo who chronicled the famous Torricellian experiment, should be appreciated in light of their contributions to pneumatics. Transforming the notion of a weightless atmosphere into an "ocean of air"—a heavy body that exerts pressure in every direction—these mid-century developments in pneumatics are a vital yet unappreciated source of Milton's ecological poetics. The chapter explains the potential transmission of pneumatic concepts into Milton's purview via his travels to Italy and ties to Robert Boyle and explores their effect on the portrayal of the fallen atmosphere in his late masterpieces.

Chapter 6, "'Throttled at Length in the Air': Environmental Warfare and Climate Regained," examines Milton's final masterpiece *Paradise Regained*, which he published alongside *Samson Agonistes* in 1671. In reframing the central conflict in environmental terms, this chapter challenges the prevailing interpretation of the confrontation between Jesus and Satan as an intellectual battle. It argues that Jesus embodies a Pauline-Augustinian model of heroism defined around vanquishing the evil powers of the air in their own territory and Satan leans on the power afforded him over the air (as well as the other elements, fire, water, and earth) in an attempt to "environ" his opponent and wear him down. The chapter reads the environmental power that Satan leverages against the Son in the context of England's fuel crisis and the subsequent ecologically disastrous shift to coal. The cultural association of demonism and satanic weather-making with the mineral fuel powering Restoration London and producing its notorious air pollution informs the double emphasis in this brief epic—best exemplified by the Hercules-Antaeus simile—on the power Satan simultaneously strips from the earth and wields in the air. By centering the majority of temptations around food and fuel security, material spoils, and demonic weather, and situating the climactic temptation high above the earth (and thus, deep within Satan's usurped territory), Milton's brief epic contends powerfully and decisively that the gravest obstacle to human salvation is the spiritual and physical corruption of the climate. Finally, this chapter proposes an explanation from pneumatics as an illuminating analogue to Christ's triumphant stand on the temple spire and in so doing clarifies much debated questions about the miraculous nature of the feat and Milton's confidence in the experimental philosophy as guide to the truth.

CHAPTER 1

"Infant Cries"
Meteorological Voices in the Nativity Ode

Before visiting the Continent and writing his first tracts on the English church and divorce, Milton composed several poems with little apparent bearing on religious politics or civic affairs. Unpopular at Cambridge, self-excluded from the clergy, and passing several years in his parents' country homes in study, the young Milton cultivated few of the traditional ties to life in this world. Neither the political impulse "to defend & be usefull to his freinds [*sic*]," "the desire of house & family," nor any craving for "honour & repute," could induce him to take orders and disturb his commitment to study.[1] Rather he hearkened to sounds and voices from a higher plane and responded in kind with poetry that aspires to divine harmony and celebrates matters supramundane. Early works such as *Ad Patrem* (To my father), *On The Morning of Christ's Nativity* (*Nativity Ode*), *Upon the Circumcision,* and *The Passion*, which were neither commissioned nor occasioned by a person's death and thus give us a sense of Milton's own inventive tendencies at this time, are especially oriented toward cosmic subjects.[2] Indeed, Milton confesses in *At a Vacation Exercise,* a short verse oration from his college days, that "if [he] were to choose," he would write poetry about the heavens, the air, and the beginning of time (29–52). In addition to surveying the distant regions where poetry transports the soul, these early writings hold up the concept of the *musica mundana,* the unheard music of the spheres, as the ideal pattern for artistic expression.

Milton considers the intellectual value of the *musica mundana* in one of his orations at Cambridge, *Prolusion II* ("On the Harmony of the Spheres"). Without placing too much weight on its position, which by the custom of such orations would have been predetermined, we will look briefly at its conception of sound and hearing. The speech is not so much a defense of the harmony of the heavens, as its title suggests, but of a historical person, namely, the philosopher Pythagoras with whom the theory was supposed to originate. The oration begins by denying that Pythagoras believed the spheres *literally* make music, as Aristotle charges in *De Caelo,* proposing rather, that if he entertained any notion of celestial song, it was a metaphor for the spheres' lawful, continuous, and orderly motions (*CPW* 1:236).³ It also, somewhat contradictorily, concedes the probability of the legend that Pythagoras was the only mortal ever to hear the celestial music, citing his worthiness "to hold converse with the gods themselves, whose like he was, and to partake of the fellowship of heaven" (*CPW* 1:238). The reason given for why others cannot hear the sound is that we lack souls like Pythagoras's, which was "pure, chaste, and white as snow" (*CPW* 1:239). If we only possessed this inner perfection "our ears would ring and be filled with that exquisite music of the stars" (*CPW* 1:239). This physiologically vivid account of how exceptionally pure souls perceive cosmic truths troubles the oration's initial assurance that sphere music is not real. The image of ears ringing out and filling up with music urges us to conceive of the divine conversation in aural terms.

Often Milton's early poetry broadcasts cosmic and metaphysical "conversations," like the kind Pythagoras could participate in, and excludes unworthy or earthly voices. During his Cambridge days, Milton, as David Masson notes, "was conscious of '*os magna soniturum,*' the mouth formed for great utterances, so all that he does utter has a certain character and form of magnitude. The stars, the gods, time, space, Jove, immortality . . .—what are they but the intellectual commonplaces of young Milton, the phrases which his voice most fondly rolls . . . ?"⁴ In *Ad Patrem,* an elegy written in thanks and supplication for his father's approval, Milton defends his chosen craft, "divine poetry," and looks forward to a time when he and his musically gifted father will wed "our sweet songs to the smooth-voiced strings, and the stars and the vaults of both the hemispheres will make their music in reply" (159). Transcending its life on earth, Milton's poetry is fated to unite with the divine music of heaven and speak to the universe itself.

NATURAL PHILOSOPHY, OR, VOICE TRAINING FOR THE COSMIC POET

A poet intent on singing to and about the "high Platonic sphere," as Masson puts it, must tune his soul to the heavenly *harmonia.* The notion that souls could resonate with sphere music derives from the assumption in classical thought that macrocosmic and microcosmic orders were metaphysically connected. Macrobius's well-known commentary on Cicero's dream of Scipio, the locus classicus of the myth of the music of the spheres, proposes that earthly music can awaken a memory of the soul's celestial birthplace and the divine music it once knew.[5] That the soul came from a divine place and occasionally rekindles a connection with it is an ancient notion, embraced by the Neoplatonists and affirmed by the young Milton, as evidenced by *Prolusion VII* ("Learning brings more Blessings to Men than Ignorance") where he states "it is, I think, a belief familiar and generally accepted that the great Creator of the world, while constituting all else fleeting and perishable, infused into man, besides what was mortal, a certain divine spirit, a part of Himself, as it were, which is immortal, imperishable, and exempt from death and extinction" (*CPW* 1:291).[6] While he would later reverse his views on the immortality of the soul, his early work was energized by the Platonic notion of a migratory soul, trapped by the body and awaiting eventual reconciliation with the divine.

Yet the super-sensitive, immortal soul that the young Milton wishes to revive in man with the harmonious sound of his poetry is not strictly immaterial, nor distinct from the natural world. The refinements *Prolusion II* claims are requisite for the soul's harmony with its heavenly source—purity, chastity, and likeness to snow—represent a more physical than mental kind of attunement. The significance of the soul's material condition in Milton's model of transcendent perception is absent from that of the Neoplatonists, who typically conceived of ecstasy or rapture as a function of dreaming or the imagination. In their view the astral "container" or vehicle that originally conveys the soul into the body flows out of it in divinatory dreams via man's *spiritus phantasticus,* or his imaginative faculties.[7] To be sure, Milton deems the imagination, as well as intellect and learning, as crucial to liberating the spirit from the body; but he also seems to regard the soul's materiality as a precondition of its activation and perception of heavenly sound. His account of perceptual *harmonia* fuses together somewhat conflicting strands of thought relating to, on the one hand, Platonic idealist cosmologies, and on the other, more materialist systems. Of the latter, Stoic physics is an apparent influence because its core concept of pneuma bridges the dualist division between body and soul.

A universally distributed substance that acts as the vital, organizing principle in all matter, pneuma is conceptually similar to the material soul-stuff that, by Milton's account, enabled Pythagoras to hear the celestial music.[8] The pneuma controlling each person's *psuchē,* and thus all their physiological and perceptual functions, persists after the body dies (as the young Milton supposed of the soul). Unlike the immortal soul of Christianized Platonism, pneuma in Stoic philosophy is a corporeal substance similar to the divine ingredient in all organisms and inanimate things, and thus fundamental to cosmology as well as psychology.[9] Etymologically speaking, this vital element in Stoic thought, which connects the individual soul to that of the world, is identified with organic and earthly vapors. Outside of philosophy the noun pneuma simply denotes "breath" or "breeze" and derives from the verb "to blow."[10] In Milton's work, the mechanisms through which the well-tempered soul communicates with the divine are pneuma-like in the sense that they, too, are both oral and ecological. Pythagoras's soul rings with the celestial song partly because of its meteorological composition: it is "white as snow."

In place of the mystical pneuma, Milton designates the atmosphere in its variety of forms—snow, cloud, vapor, thunder, lightning, etc.—as the conduit to transcendent vocal expression. Situated between the celestial orbs and the human realm, it plays a pivotal role in Milton's understanding of what we might call the great sonic chain of being. Air is the animate medium capacitating aural exchange between mundane and heavenly spheres. In *At a Vacation Exercise,* written not long before the *Nativity Ode* and delivered at a college entertainment, Milton contemplates what he might speak about if the choice were his. To communicate this "graver subject," he explains, would require "fit sound" (30, 32). The quest for sound worthy of his theme takes place in the middle part of the oration, where the speaker pursues his fancy into the far reaches of the cosmos. In the heights of heaven the speaker's actions are noiseless; all he can do is look at the blissful gods and listen to Apollo's music, possibly with the aim of emulating it. To speak, however, it seems he must descend

> through the spheres of watchful fire,
> And misty regions of wide air next under,
> And hills of snow and lofts of piled thunder.
> (40–42)

Only after traversing these layers of atmosphere may he "tell at length" and "sing" the secrets of the world (43, 45). The mental trip the poet takes before speaking (the opening part of *At a Vacation Exercise* depicts a similar but more literal process of "clearing the throat") demonstrates that atmosphere is integral to both the physical act of speaking—supplying the wind in his lungs and allowing him to vocalize his divinely inspired tune—as well as poetic invention. He may derive the pattern of his harmony from "[a]bove the wheeling poles," but to fashion and inspire his own song he needs the native element of air (34). The polysyndeton in the lines depicting the speaker's passage through the skies ("And misty regions . . . / And hills . . . and lofts") suggests that by plummeting through the clouds he gains a kind of imaginative momentum.

While the air can give substance to the heavenly sounds that the soul perceives in its flights of fancy, often in Milton's poetry atmosphere poses an obstruction or hazard to the clear perception of divine sounds. Though it may be channeled to record and resonate the heavenly harmonia, atmosphere is also a medium that has been corrupted by sin. To remind readers that atmosphere in its corrupted state is the basis of all earthly utterance, Milton emphasizes environmental voices in lieu of human ones in the early poetry. In underlining the vocal shape of the climate and the environmental properties of the voice, Milton brings to light the dependence of speech and poetic expression on an element that is potentially generative of both divine and corrupt sounds. The air poses both a pathway and stumbling block to the poet whose aim is to hear the divine music and make it intelligible to others.

Since all earthly voices stem from the atmosphere and through it some aspire to converse with the divine, Milton defines certain avenues for correctly harnessing the vocal power of the air. To cast one's voice into the divine conversation, for example, one must possess deep knowledge of Creation. In *Prolusion VII*, Milton asserts "a man who is almost entirely absorbed and immersed in study finds it much easier to converse with gods than with men" (*CPW* 1:295). If too much study diminishes one's interpersonal skills, then its reward is a stronger voice for honoring God: "The great Artificer of this mighty fabric established it for His own glory. The more deeply we delve into the wondrous wisdom, the marvelous skill, and the astounding variety of its creation (which we cannot do without the aid of Learning), the greater grows the wonder and awe we feel for its Creator and the louder the praises we offer Him" (*CPW* 1:291–92). After learning everything there is to know about the world, "the spirit of man, no longer confined within this dark prison-house, will reach out far and wide, till it fills the whole world and the space far beyond with the expansion of its divine greatness" (*CPW* 1:296). The human voice is upborne with the soul and invested with power over the forces of Nature. He who understands the natural world and can predict its vicissitudes "will indeed seem to be one whose rule and dominion the stars obey, to whose command earth and sea hearken, and whom winds and tempests serve" (*CPW* 1:296). He speaks to these entities not through apostrophe, the mode usually applied to inanimate things, but dialogically with the expectation they will comprehend and obey him.

While Milton's juvenilia do not name the specific kinds of philosophy man should pursue in order to know and command Nature, they imply the importance of natural sciences in general. In the scenario depicted below from *Prolusion VII*, man's complete mastery of natural law depends on his combined knowledge of several branches of philosophy including astronomy, physics, and metaphysics:

> What a thing it is to grasp the nature of the whole firmament and its stars, all the movements and changes of the atmosphere, whether it strikes terror into ignorant minds by the majestic roll of thunder or by fiery comets, or whether it freezes into snow or hail, or whether again it falls softly and gently in showers of dew; then perfectly to understand the shifting winds and all the exhalations and vapours which earth and sea give forth; next to know the hidden virtues of plants and metals and understand the nature and the feelings, if that may be, of every living creature; next the delicate structure of the human body and the art of keeping it in health; and, to crown all, the divine might and power of the soul, and any knowledge we may have gained governing those beings which we call spirits and genii and daemons. (*CPW* 1:295–96)

Though this description privileges wisdom in matters of the soul, it also exalts knowledge about all aspects of the cosmos. "Universal learning" is necessary for achieving intellectual rapture—the spirit's release from the "dark prison-house" of the body into the space beyond the world (*CPW* 1.296). Almost Lucretian in its rationalism, this educational program promises to dispel distressing uncertainty about the world. By dwelling on the variety and potentially frightening nature of meteorological activities, Milton implies that the vacillations of the atmosphere are a major source of this uncertainty. In comprehending the complexity of the weather, which "[strikes] terror into ignorant minds," we overcome our greatest fears of the unknown.

Milton's emphasis on meteorological phenomena in this passage possibly reflects the fact that a similar plan for "universal learning" appears at the beginning of Aristotle's *Meteorologica* and includes an extended account of the sublunary atmosphere and its changes. Though the curriculum Milton envisions in *Prolusion VII* covers a few topics that Aristotle's summary leaves out (namely, medicine and metaphysics), in large part it resembles Aristotle's rubric, which delineates, probably for the first time, the scope of natural philosophy and all its branches.[11] Aristotle, like Milton after him, follows a spatial organization in his précis of natural philosophy, beginning with the macrocosm and proceeding to the microcosm. Celestial subjects are treated first, atmospheric movement next, and the study of plants and animals last.[12] In *Ad Patrem*, Milton provides a similar map of the divisions of science when he expresses gratitude for the chance to learn about all that exists "in the sky, or on mother earth beneath the sky, or in the air that streams between them, or hidden beneath the waves, beneath the heaving marbly surface of the ocean" (160). The space we now refer to as the "atmosphere" is neither vacuous nor uncomplicated in this description. Air constitutes its own middle region between the sky and the earth, and its phenomena warrant study in their own right.

Given that epistemology in these examples is conceived in spatial terms, with cosmography determining disciplinary boundaries, it is not surprising that Milton frequently emphasizes the environmental context of aural phenomena in his early poetry. Because natural objects are known in part by their location in the cosmic hierarchy, the place from which sound originates and through which it permeates is perhaps its most meaningful attribute. The opening lines from *The Passion*, which allude to the *Nativity Ode*'s depiction of heavenly music on earth, use a theater conceit to construe the songs of the angelic choirs, and indeed Milton's ode itself, as fleeting: "Erewhile of music, and ethereal mirth, / Where with the stage of air and earth did ring" (1–2). The interval in which such music may be sustained on earth is short, as brief as a play. In *Ad Patrem* sonic environments can be revealing even if the sounds themselves are not audible. The only way of imagining the music of the afterlife, for instance, a "deathless melody, an indescribable song," is to contemplate the celestial place where the spirit will ascend (and where Milton claims poetry has already carried his spirit) "round the hurtling spheres" and "among the starry choirs" (159). Similarly, in the *Nativity Ode* Milton refers to earthly voices metonymically by way of the atmosphere and uses the hierarchy of cosmic space to indicate the inferiority of atmospheric voices to heavenly sounds but also the potential for mutual exchange between human and divine realms.

Outside of his poetry, we do not have biographical evidence that Milton studied the atmosphere or was expert in meteorological theory. Nevertheless, this might well have been the case. At Cambridge he likely encountered Aristotle's theory of meteors, either directly or in a Scholastic textbook like Bartholomaeus Keckermann's *Systema Systematum* (1613), which includes a section on meteoric phenomena and their causes.[13] Milton makes an explicit reference to the *Meteorologica* in *Prolusion II* where he alludes to Aristotle's discussion of "goats" or shooting stars (*CPW* 1:238).[14] After his university days, Milton probably continued to encounter meteorological writings in his independent studies. In 1631, he purchased a copy of Aratus's *Phaenomena* (Paris, 1559), a classical Greek poem whose latter part offers interpretations of weather signs found in astronomy and wildlife. Some of the annotations in this volume are from Milton's hand and thought to originate from two different periods in his life—directly after he bought the book and later, when he tutored his nephews.[15] Though most of the notes refer to grammatical issues, one passage that Milton marked pertains to weather prediction and seems to have inspired similar verses in *Paradise Lost*.[16] Since the *Phaenomena* is one of just a few extant volumes containing Milton's notes (many others must have existed), it does not tell us the extent to which he read meteorological lore. But its annotations prove that he had engaged closely with a literary text about weather signification, a tradition that influences his depiction of meteorological phenomena as communicating messages of a divine origin.[17]

The curriculum Milton outlines in *Of Education* (1644) exposes pupils to the "history of meteors" after they have mastered Latin and Greek (*CPW* 2:392). Since Greek was necessary to advance to this subject, we may assume that Milton intended for students to learn meteorology from the classics, possibly in tandem with Renaissance compendiums of science. If the educational program this tract espouses resembles at all the kind of reading Milton himself undertook, either at Cambridge or his father's country home in Horton, where, as he states in *The Second Defense of the English People,* he devoted himself "entirely to the study of Greek and Latin writers," then it is probable that his knowledge of meteorology was founded on the classic works of natural philosophy (*CPW* 4.1:614). The major texts containing explanations of weather besides Aristotle's *Meteorologica* are Seneca's *Naturales quaestiones* (*Natural Questions*), Pliny the Elder's *Naturalis historia* (*Natural History*), and Lucretius's *De rerum natura* (*On the Nature of Things*).

As I have argued, Milton considers natural philosophy as a means of gaining control over Nature and endowing the voice with authority verging on that of the divine. He views the science of meteorology as particularly efficacious in bestowing this kind of authority. That practical knowledge of meteorological phenomena may refine and empower the human voice perhaps seems nonsensical. But reasons for linking the weather with speech and sound—its clarity, audibility, or potency—may be found in the language of the meteorological tradition and in Milton's persistent identification of the voice with spaces above and beyond the domestic or mundane. The next section argues that Milton associates the earth's aptitude for producing sounds—speech, noise, and music, for instance—with the atmosphere, its motions, and inhabitants rather than human activities, and ventures an explanation for why he emphasizes the environmental aspect of acoustics in his early poems. I submit that the meteorological tradition with its rich oral and acoustic vocabulary for theorizing natural processes underlies Milton's conception of the voice in the *Nativity Ode.*

In December 1629, Milton wrote in "Elegia sexta [Elegy VI]" to Charles Diodati about his latest project, a Christmas poem: "I am writing a poem about the king who was born of heavenly seed, and who brought peace to men.

I am writing about the blessed ages promised in Holy Scripture, about the infant cries of God, about the stabling under a poor roof of Him who dwells with his Father in the highest heavens, about the sky's giving birth to a new star, about the hosts who sang in the air, and about the pagan gods suddenly shattered in their own shrines" (122). This description of the *Nativity Ode* is remarkably focused on the setting of the holy occasion, enumerating all of the cosmic spaces that are implicated by Christ's birth: a low stable shelters a king who hails from the "highest heavens"; the sky boasts a new star; the air is filled with divine music; and the earth's pagan haunts become disenchanted all at once. In illustrating the effect of Christ's nativity radiating through distinct segments of the cosmos, the summary is reminiscent of the accounts of the natural sciences in *Prolusion VII* and *Ad Patrem*. This is not surprising. The dynamics of cosmography—the interrelation between the regions of heaven, air, and earth, the porousness of their boundaries, and the movement of beings between them—are central concerns of Milton's early works and remain integral to his later poetry.

As much as these dynamics serve to organize and hierarchize knowledge of the universe, illustrating the intricately connected design of creation, they also produce an expansive, impersonal perspective in poems like the *Nativity Ode*. Rather than dwell on the animals by the manger, the shepherds and the magi, or Mary and Joseph, Milton's ode marginalizes the characters traditionally at the center of the nativity story. As J. Martin Evans has argued, "The entire scene, one could say without exaggeration, has been completely dehumanized."[18] Whereas depictions of the human community that received Christ might emphasize his humanity, the poem's attentiveness to how the environment responds to and receives inhabitants from the higher spheres stresses the relationship between heaven and earth and brings out the savior's qualities as a ruler and redeemer: he comes to Earth to triumph over man's enemies and assume his role as humanity's awful judge.

The *Nativity Ode* reverses Milton's situation in *At a Vacation Exercise* where his own babbling "infant lips" must learn to speak and seek "fit sound" for a journey to heaven and back. The ode, by contrast, depicts God's "infant cries" as immediately fit to quell the strife on earth. Yet in its fully realized form, the poem never represents the Christ child's voice. It relies on an assortment of nonhuman sounds as well as intervals of silence to announce Jesus's presence. The poem thus illustrates Milton's association of the high heroic voice, perfected in the divine Logos, with commanding power over cosmic spaces, their motions, and affiliations.

In the *Nativity Ode* the atmosphere is a morally ambiguous acoustic space which, when moved by music or poetry, can be made to sympathize with the heavenly harmonia but is also easily contaminated or manipulated by corrupt forces. Accordingly, the poem represents the air as an undecided zone, a region courted by earthly and ethereal agents and almost conquered at the nativity by the irresistible presence of God. Though the poem looks forward to and imagines a future time when the air might be converted to making perfect sounds, Milton cannot depict the atmosphere as already conquered and subsumed into heaven at the Incarnation. Instead, he portrays the air as a compromised and yielding substance compelled to answer Nature's law though preferring the sweet music of heaven. Milton stages the birth of Christ as an acoustical contest over the air because the Incarnation enacts the first phase of a kind of sanctification of the universal voice. At the moment of Christ's birth, when the Word came into the world, it did not simply speak to a few; it seized the entire atmosphere so that it might express itself through Nature and eliminate false influences from every voice. But the sanctification is not entire, and Milton assures us that some evil remains in the air.

Milton would also have had personal reasons for placing the atmosphere at the center of his poem. Since he wrote the ode around the time when he was deciding on a vocation, a possible driving motivation behind it was to better understand the moral status of the air, which would serve as the material of his poetic craft. The capacity of the air for joining in or supporting divine music, which Milton viewed as a poetic ideal, would certainly concern an earnest young poet, anxious to use his talents in the service of God. And it is this question that the *Nativity Ode* explores by imagining the kinds of sounds the air might produce or withhold in the presence of heaven's king.

CONTENDING FOR CLIMATE CONTROL

After a brief proem addressed to the heavenly Muse, the section of the ode titled "Hymn" begins by describing why Christ was born during "the winter wild" (29). The season is attributed to Nature's anxiousness to appear chaste and sober to the Lord:

> Nature in awe to him
> Had doffed her gaudy trim,
>   With her great master so to sympathize:
> It was no season then for her
> To wanton with the sun her lusty paramour.
>   (32–36)

The next stanza continues the conceit that Nature is a coquette currently on her best behavior, and even goes so far as to imply that sexual promiscuity has left her body diseased and scarred. To hide her "foul deformities" from God and paint herself a virginal hue, she covers herself with a white "saintly veil" of snow (42–44). But Nature, whose "front" or body is associated with the earth itself, is not capable of furnishing this snowy disguise on her own. She needs the air's cooperation:

> Only with speeches fair
> She woos the gentle air
>   To hide her guilty front with innocent snow.
>   (37–39)

The conceit makes the air seem both independent from Nature, but also under her power, which lies in her ability to persuade or "woo" the air with speech. The details of this interaction are somewhat enigmatic. Why does Nature use speech (a distinctly human invention) and not some other means to coax snow from the sky? What do her fair speeches consist of, or what do they signify apart from flattery? The previous stanza tells us that in warmer seasons, Nature would "wanton with the sun," implying that its warmth could be won by Nature's physical or visual appeal. But the atmosphere evidently responds to a different kind of overture, one that is elementally similar to it. It is because speech is made of air that the atmosphere *listens* to Nature's words. There is also the suggestion that Nature's verbal or vocal dalliance with the sky will be less offensive to the incarnate Lord than overt spectacle. The qualifying word "only" in the statement that Nature "[o]nly with speeches fair / . . . woos the gentle air" (emphasis added), casts her speech as a lesser evil than the "gaudy trim" she flaunts in more temperate seasons.

There are additional reasons why Milton might characterize the falling of snow as a response to Nature's spoken commands, and, moreover, why he would devote two stanzas to explaining how this historic winter came about. One very probable explanation for Milton's portrayal of Nature as speaking to the air and telling it to snow is that from antiquity most meteorological phenomena were attributed to exhalations—vapors, believed to originate from the ground and its bodies of water, which some writers described as the earth's breath. Exhalations became fundamental to meteorological theory because of their prominence in Aristotle's *Meteorologica*, where they are invoked to explain nearly every kind of weather occurrence.[19] Exhalation theory enlarges on more primitive meteorological ideas, like those of the pre-Socratic philosophers Anaximander and Anaximenes, whose views mostly survive in the works of other classical writers. Anaximander seems to have believed that pneuma shaped all meteorological activity, and for Anaximenes, who was said to be Anaximander's student, *aēr* is the primary element behind the weather.[20] Accounts of Anaximenes's philosophy indicate that he believed *aēr* is the basis not only of meteorological phenomena, but also all other forms of matter, for example, fire, wind, earth, and stone.[21] By attributing meteorological and other natural activities to transformations of *aēr* or pneuma—both ambiguously translated terms with a wide semantic range—some accounts of these Milesians' views suggest that they believed the world is animated by a universal breath or something like an individual soul.[22]

Aristotle's concept of exhalations has a narrower, more technical definition than that pertaining to Anaximander and Anaximenes's weather-causing gases. He describes two types of exhalation—one arid and hot; the other, moist and cold—that in combination constitute the air.[23] In addition to acting jointly to cause weather phenomena, each exhalation can produce its own distinct meteorological effects.[24] Although Aristotle does not claim that exhalations are the earth's respirations, or, by the same token, that the earth is alive, he sometimes employs biological metaphors to explain the movement of exhalation throughout the universe.[25] The very language of exhalation theory as used by Aristotle implies the organism-like structure of the earth. Indeed, the Greek word for exhalation, ἀναθυμίασις, applies both to geological activity and to rising vapors in bodily processes and in the soul. While descriptions of exhalations in *Meteorologica* thus substantiate a physical theory of the sublunary world rather than a theistic or vitalistic one, they nonetheless suggest that processes of evaporation are similar to respiration and even digestion.[26] Seneca's definition of wind refers to the earth as emitting breath: "Sometimes the earth itself ejects a great quantity of air, which she breathes out [*spirat*] from hidden recesses."[27] The Latin words that early modern commentators used to render Aristotle's exhalations, *exhalatio, spiratio,* and *halitus,* all stress their relationship to respiration.[28]

The *Meteorologica* touches on the phenomenon of snow only once and briefly. The main explanation Aristotle offers for it is, that "when cloud freezes snow is produced, when vapour, hoar frost."[29] He is more concerned with explaining the processes of evaporation and cloud formation, which occur continuously and account for all kinds of precipitation, including snow. According to Aristotle, clouds form when the sun draws up vapor from the earth and the heat dissipates, causing the vapor to condense into water: "The exhalation from water is vapor; the formation of water from air produces cloud."[30] Refrigeration dispels more heat from the cloud, causing it to produce snow.[31] Thus, it is the combination of moist exhalation and intense cold that brings about snow. Both factors are potentially present in Milton's depiction of winter at the beginning of the *Nativity Ode*. Nature's abstention from "wanton[ing] with the sun" possibly alludes to the remote position of the sun in the winter sky. In Aristotle, evaporation and water formation in the air depend on the sun's position in the ecliptic.[32] The nearer the sun is to the "stream of vapor" that surrounds the earth, the more moisture in the air rises; when the sun is farther from the vapor (as in the winter months) the more it cools and falls to the earth.[33] The sun's estrangement from Nature in the first stanza of Milton's poem indicates that the atmosphere is cold and ready to produce snow. By the same token, Nature's "speeches fair" are akin to the moist exhalations or vapors that are the material cause of clouds and snow.

Though Nature at first succeeds in coaxing the air to snow, her power over the air is suspended for the remainder of the poem. Christ's power over the elements, particularly manifest in the attraction of heavenly song, supersedes Nature's ability to command the air to be meteorologically productive. Both Nature and the representatives of heaven conquer the air either by compelling it to participate in acoustical acts or altering the sounds it makes. We have seen that the meteorological activity at the beginning of the *Nativity Ode* is depicted in oral terms as is often the case in classical natural philosophy. Nature's weather-causing exhalations are portrayed as speeches and the snow as the air's obedient response. With Christ's coming, however, the animating principles or law of Nature, which are manifested in the poem by its "speech," no longer hold sway over the cosmic landscape. God's speech is required to move the starry sphere, which neglects the natural law that governs its motion because of its awe for Jesus:

> The stars with deep amaze
> Stand fixed in steadfast gaze
> . . . .
> Until their Lord himself bespake, and bid them go.
> (69–76)

God's agents, the "meek-eyed Peace" and the "helmed cherubim / And sworded seraphim," extend this divine authority into the sublunary realm, establishing their rule by quelling discordant sounds and replacing them with exquisite, heavenly music (46, 112–13).

The acoustical method that the divine agents use to subdue and convert the earth has a particularly powerful effect on the atmosphere. Milton repeatedly describes the air and its phenomena as being enamored with the heavenly host. This love or infatuation with the divine presence manifests itself in two different ways. At first the air falls into a state of silent captivation, and later, is moved to harmonize with the divine music. When God sends Peace into the world, she comes "crowned with olive green" and

>   softly sliding
> Down through the turning sphere
> . . . .
>   With turtle wing the amorous clouds dividing.
>   (47–50)

The clouds are "amorous" towards her; she moves through them as a queen through crowds of admiring subjects. Once the "reign of peace" is established, "[n]o war, or battle's sound / Was heard the world around," and a hush comes over Nature:

> The winds with wonder whist,
> Smoothly the waters kissed,
>   Whispering new joys to the mild ocean.
>   (53–54, 63, 64–66)

The taming of Nature and subsidence of dissonant war mark the first stage in the world's aural renewal. Only in the absence of such noise may the air express the refined sounds of heaven. After this lull comes over the land and the heavenly host breaks into song, the air's silent admiration thus turns to ecstasy. Their music sends the shepherds' souls into rapture, and "[t]he air such pleasure loth to lose, / With thousand echoes still prolongs each heavenly close" (98–100). In contrast to the earlier passage where the air is wooed by Nature's speech and consequently produces snow, here it is courted by a "divinely-warbled voice," and responds by making echoes (96). These responses differ in purpose as well as in kind. Whereas the snow at the beginning of the poem is meant to mitigate Nature's shame by covering her up, these echoes make the divine music more audible and permanent. Not content to be a mere passive medium, the atmosphere participates in reproducing and preserving the sound of the angels' otherworldly music. The air's desire to luxuriate in and extend the angels' song, along with the fact that the stars refuse to revolve except at God's command, indicate that the environment is more strongly compelled by divine voices and sounds than by natural ones.

When she hears the rapturous sound of the angels' music, Nature herself concludes that the world, chiefly the air, prefers the melody of heaven to her own spoken commandments:

> Nature that heard such sound
> Beneath the hollow round
>   Of Cynthia's seat, the airy region thrilling,
> Now was almost won
> To think her part was done,
>   And that her reign had here its last fulfilling:
> She knew such harmony alone
> Could hold all heaven and earth in happier union.
>   (101–8)

Nature, the wooer, is herself "almost won"; in other words, she is almost convinced that her power over the physical world has been discontinued and replaced by a ruler capable of offering what she cannot: bliss and unity for all of creation. Nature's temporary defeat is evinced by the fact that heaven's superior harmony now possesses the "airy region" where Nature's speeches formerly held sway.

The stanza elaborates one of the *Nativity Ode*'s major themes by presenting heaven and Nature (or earth) as rival monarchs jockeying for sovereignty over the air. It foreshadows, for instance, the images of conquest in stanza seventeen that look forward to the Second Coming when "the dreadful judge *in middle air* shall spread his throne" (164; emphasis added). The seventeenth stanza more fully realizes the eschatology hinted at in stanza ten (above) by depicting heaven's actual annexation of the air as well as the real overthrow of Nature at the end of time. Though she judges prematurely in stanza ten that "her part was done," Nature's intuition that the apocalypse will be heard, as much as seen or felt, is apt. Like the nativity, the Last Judgment is presented as aural event that takes effect in the atmosphere and underground (where philosophers believed winds and exhalations were also present).[34] Hence the admonishment given by "wisest fate" that the world will not be truly redeemed until the "trump of doom" sounds

> With such a horrid clang
> As on Mount Sinai rang
>   While the red fire, and smould'ring clouds out brake:
> The aged earth aghast
> With terror of that blast,
>   Shall from the surface to the centre shake.
>   (156–62)

According to this description, the world will be destabilized rather than unified by the sound of the apocalypse, which summons clouds and fire to the sky and undermines the foundations of the earth. Before the harmony of divine music joins together heaven and earth eternally, the corrupted voice of Nature—its vital exhalations—must be blasted from the deep crevices of the earth. Alluding typologically to the music of the nativity, and before that, the trumpet announcing God's deliverance of the Law on quaking, clouded Mount Sinai, this passage represents the Lord's final triumph over the air and the complete conversion of the atmosphere into an instrument of heaven. Clouds appear not because Nature calls for them—she has no power in "the world's last session"—but because they manifest God's might (163).

The notion of the atmosphere as a zone of contention between forces from above and below was a familiar one from Pliny's *Natural History,* which contains sections on meteorological causes and weather prediction. Pliny's vivid description of the terrestrial atmosphere portrays the air as continually compressed and pulled on from opposite directions, and as a consequence, swept over with the meteorological phenomena generated by these motions:

> The force of the stars presses down terrestrial objects that strive to move towards the sky, and also draws to itself things that lack spontaneous levitation. Rain falls, clouds rise, rivers dry up, hailstorms sweep down; rays scorch, and impinging from every side on the earth in the middle of the world, then are broken and recoil and carry with them the moisture they have drunk up. . . . Empty winds sweep down, and then go back again with their plunder. So many living creatures draw their breath from the upper air; but the air strives in the opposite direction, and the earth pours back breath to the sky as if to a vacuum. Thus as nature swings to and fro like a kind of sling, discord is kindled by the velocity of the world's motion. Nor is the battle allowed to stand still, but is continually carried up and whirled round, displaying in an immense globe that encircles the world the causes of things, continually overspreading another and another heaven interwoven with the clouds.[35]

Whereas a dizzying number of astronomical and meteorological forces are mentioned in this passage, a few main causes are in play: the downward and upward pull of the celestial bodies, the earth's outward and inward breaths or exhalations, and the circular motion of the heavens. The first two causes are reciprocal, accounting for the "to and fro" of vapors, winds, and precipitation between the earth and sky.[36] The latter cause, the horizontal sweeping motion due to the continuous revolution of the celestial orbs, infuses "discord" into the "battle" that Pliny stages between the heavens and the earth. Descriptions of marauding winds and sunbeams impinging on or striking against (*impellunt*) the earth reinforce the idea that the land has a sometimes-hostile relationship with the regions overhead.

Pliny also discusses the push-and-pull of the stars alluded to in the passage above elsewhere in the *Natural History*. He seems to think of the world as held together by the stratification of elements that have naturally opposed weights: "Thus the mutual embrace of the unlike results in an intertlacing [sic], the light substances being prevented by the heavy ones from flying up, while on the contrary the heavy substances are held from crashing down by the upward tendency of the light ones."[37] The planets therefore are "upheld by the same vapor [air] between earth and heaven."[38] Though they are buoyed up by the air, the stars nevertheless cast down various kinds of meteorological phenomena into the atmosphere: "Some [stars] are productive of moisture dissolved into liquid, others of moisture hardened into frost or coagulated into snow or frozen into hail, others of a blast of air, others of warmth or heat, others of dew, others of cold."[39] For Pliny the stars and planets supply the "regular" causes of meteorological turbulence as opposed to the earth's upward-tending vapors, whose atmospheric effects are merely "accidental."[40] Thunderbolts, for instance, have various causes, some of them regular and some accidental. They can also arise from a mixed cause, the product of a kind of battle between the terrestrial exhalation and the downward-tending pressure of the stars: "It is also possible for breath emerging from the earth, when pressed down by the counter-impact of the stars, to be checked by a cloud and so cause thunder, nature choking down the sound while the struggle goes on but the crash sounding when the breath bursts out, as when a skin is stretched by being blown into."[41] The thrust of the heavenly bodies causes Nature to choke down the sound of its breath, leading eventually to an acoustically dramatic explosion.

Yet the contest for the air depicted in the *Nativity Ode* differs somewhat from the battle endlessly waged in Pliny's atmosphere by the motions of the terrestrial and celestial realms. At the birth of Christ, heaven's agents charm the atmosphere without using any violence. Peace is sent into the world to calm Nature's fears, and though she "strikes" the earth with her wand, its blow sends out a universal peace, not discord (51–52). Her arrival momentarily repairs the climate disfigurement depicted in the episode of *Paradise Lost* in which the angels disrupt the perfect functioning of the universe and "Outrage" and "Discord" appear among the earth's elements and creatures (10.706). In the *Nativity Ode*, God's presence in the world quiets the winds, which act thereafter as agents of unity rather than disunity between Nature's regions. Far from stirring up turbulence in the atmosphere, the stars focus their "precious influence" (instead of more harmful emissions) exclusively on the newborn Lord (71). All of these signs indicate that heaven's encroachment on the atmosphere at the nativity is welcome. Even so, the political language of the conceit, combined with the fact that Nature expresses fears about a possible ouster, inevitably call to mind invasions of a more forceful nature. The ode's allusion to the thundering sound of the Last Judgment that will rock the earth to its very core, moreover, testifies that God's reign in the sky will not always be peaceful. At an uncertain future date heaven will harness the air to inflict a terrible punishment on the earth.

The poem integrates the Plinian model of the atmosphere as a warzone by depicting Nature's "front" (earth) and heaven's agents as rivals in a contest over the air. As in Pliny's scheme, the earth's rising breaths have acoustical and meteorological potency, commanding snow to fall from the air with "speeches fair." Their power to affect the atmosphere's sounds and its weather is offset by forces from above that God endows with even greater influence over the air. This scenario mirrors the stars' counteraction of the earth's exhalations in Pliny's *Natural History*. The celestial bodies' meteorological efficacy seems vaster than that of the earth, perhaps because the influence of the stars is not purely accidental but "prophetical and sent from on high."[42]

"TO HIS CELESTIAL CONSORT US UNITE": NATURE'S BINDING AIRS, BREATH, AND SWEAT

Nature's awareness on hearing the divine sound of the angelic choir that "such harmony alone / Could hold all heaven and earth in happier union" (107–8), offers further insight into Milton's views about the cosmic role of sound. Her knowledge implies that before the Incarnation (nearly) ushers in a Golden Age with its miraculous harmonies, the unity of the world was in her charge. By reminding us that the healing concord of the spheres cannot join heaven to earth until Christ "redeem[s] our loss," the ode also suggests that Nature will resume her station as a unifier *after* the nativity and until the Second Coming, perhaps in a modified capacity (125–56). How does Nature play her "part" in the great cosmic song? By now it should be clear that Milton conceives of her shaping influence on the atmosphere, at least, in terms of a universal voice. The global climate is dictated by what either Nature or God says to the air. Yet the spoken instructions Nature gives to the elements in the *Nativity Ode* are inferior to the music of the angel choir, which achieves the perfect union of divine voice and instrument. The implication is that Nature's aural-meteorological expressions have a less powerful effect on the world than the rapturous sound of the nativity, which represents the initial phase of a Providential plan to renew and replace the harmony that binds the universe together.[43]

The idea that music possesses a kind of cosmic binding power was not new. The notion has its roots in the Greek concept of harmonia, which could refer to the order in the cosmos as well as the proportions found in musical harmonies or scales.[44] Based on the Pythagoreans' discovery that the most mathematically "perfect" ratios correspond to pleasing sounding intervals in Greek musical scales, they presumed that these numbers also govern the structure and motions of the entire cosmos. Along these lines Western thinkers up through the Renaissance, notably Plato and Boethius, used musical metaphors to explain the harmonious motions of the spheres and the metaphysical relation between the heavens, the elements, man, and even aspects of civilization.[45] Related to the idea of world harmony is the tradition, rooted in the classical philosophy of nature and Hermetic writings, that the breath of the world—pneuma or spirit—exhales itself in the form of music.[46] Underlying this link between spirit and music is the Pythagorean myth of the spheres and its afterlife in Western philosophy: "World spirit, because it is the breath of a musically proportioned universe, is musical. Conversely, music, because it images cosmic music, possesses or is spirit."[47] It was primarily the philosophy of Marsilio Ficino that cemented this relationship by positing that the universe is alive, infused with spirit similar to music, and further, that its parts might be accessed and drawn into each other through song.[48] Earlier formulations gave pneuma or world spirit another defining attribute, similar to the metamusical concept of harmonia, but more general. For the Stoics, pneuma has a kind of holding power that gives unified objects their natural shape by drawing to the center all of their parts: "As the active ingredient in all things in the world, pneuma is responsible for the 'tension' that holds all the world and everything in it together."[49] The metaphor of world music or harmony, then, accommodated not only the breath-like aspect of pneuma, but also its role as the glue that actively holds the universe together.

Milton uses this metaphor to express the divine music's capacity, like that of pneuma, to bring together or heal the distance between heaven and earth and simultaneously imprison any forces of opposition within its circumference. "For if such holy song / Enwrap our fancy long," the speaker muses, the Golden Age will come again and "heaven as at some festival, / Will open wide the gates of her high palace hall" (133–34, 147–48). The binding potential of heaven's sound is also realized beneath the ground. If the heavenly music were to linger on earth "hell itself will pass away," but even its brief presence places strictures on "[t]he old dragon under ground," who finds himself at the moment of Christ's birth "[i]n straiter limits bound" (139, 168–69). The formidable Christ child, who, like a new sun conquers the pagan gods as if with planetary spirit, the blinding "rays of Bethlehem," is also armed with the power of constriction: "Our babe to show his Godhead true, / Can in his swaddling bands control the damned crew" (223, 227–28). The son's ability to hold together, bind, or unite the parts of the earth comes a boon to Creation and a bane to its enemies.

The ode's representations of divine song—the "music sweet" of the angel choir, the imagined accompaniment of the crystal spheres, and the recollected song of Creation—imagine a sound so exquisite and so compelling that no cosmic flaw can withstand the fortifying restorative process it initiates in the world. But they also describe either exceptional moments in time or an ideal reality when Nature is either unfallen or divinely regenerated. Barring this, the earth's atmosphere is incapable of producing such alleviating music on its own. Without the informing influence of the Word or Logos, Nature lacks the score of the divine song.

The promise of a "happier union" that Nature perceives in the divine sounds of the angelic choir suggests that there are two possible versions of cosmic embrace, one metamusical and one physical. We have already touched on the first, which is world harmony, or a state of cosmic order that reflects and expresses the proportioned, mathematical nature of music. The choir's euphony in the *Nativity Ode* is an emblem of what Milton considers the consummate form of world harmony, which can only be attained through the redemption of Christ, and it also represents the actual vehicle of this unification: the infusion of the air with music, that life-giving spirit that knits together the regions of heaven and earth.[50] The other version of world synthesis glanced at in Nature's allusion to a "happier union" is, of course, the *less* happy form of unity that Nature offers prior to the coming of Christ and must sustain until his return. Recalling Pliny's depiction of the atmosphere from the *Natural History* may help us to imagine what Nature's imperfect hold on the world resembles. In that description, moisture gets pulled up and buffeted down again, and the sky continually sucks the breath from the earth, as do its creatures from the sky. The layers of elements in between—earth, water, air, and fire—are firmly held together, implicitly by the "tension" of world spirit or breath and explicitly by opposite attractive and propulsive forces, but the result is more like a battle between the elements than a peaceful embrace. Perhaps this turbulent equilibrium is the only kind of "union" that Nature's voice may secure in the absence of the harmonizing power of the Word.

The loudness of a warring atmosphere (e.g., the raving of the ocean [67]) is not the only harsh sound supposed to have preceded the silence of that holy night. The manmade noise of war or "battle's sound" is also referenced at the beginning of the ode (53). The source of these discordances and, more broadly speaking, the weak holding power of Nature's breath remains unexplained. For answers we must turn to the last third of the hymn (stanzas 19–27), which switches from considering the miraculous universal effects of heavenly sound to examining the compromised condition of the fallen atmosphere. The necessity of including these stanzas, which describe an episode from Prudentius on the cessation of the oracles, lies in the exigencies of cosmic renewal. To make the earth anew, God needs to begin with a clean slate: "Before it can be redeemed, the world must first be purged of its impurities in a global act of exorcism."[51] It is worth noting that the evidence Milton gives of this great act of purgation is overwhelmingly oral or acoustical in nature. The departure of the pagan gods from their shrines, the earth's response to their surrender, and their worshippers' futile initiation rites are all figured in acoustical terms or with respect to the absence of sound that the oracles leave behind. These two representational strategies are apparent in the introductory stanza of this section:

> The oracles are dumb,
> No voice or hideous hum
>   Runs through the arched roof in words deceiving.
> Apollo from his shrine
> Can no more divine,
>   With hollow shriek the steep of Delphos leaving.
> No nightly trance, or breathed spell,
> Inspires the pale-eyed priest from the prophetic cell.
>
>   (173–80)

By portraying the expiration of divinatory voices the stanza reverberates, in spite of itself, with the breathy sound of these profane utterances: the "hideous hum," "words deceiving," "nightly trance," and "breathed spell." Its three "nos" symbolize the muting of the oracles; but as a mimetic device for representing silence, the negations are not quite successful. Depending on how one scans lines 174 and 179, the two initial negations, but the second one especially, receive more emphasis than a regular unstressed syllable because of their position at the head of the line and because they create a refrain. Thus, even as the word "no" semantically cancels out the sound of inspiring gods, experientially, it amplifies the sounds of the stanza. The line "[w]ith hollow shriek the steep of Delphos leaving"—the only description of an actual sound as opposed to the memory of one—confirms the impression that the other lines give: the cessation of the oracles is an audible event with profound aural consequences (178). The result is not merely the erasure of deceitful influences from the world, though this is certainly one of its effects, but also the moral and physical transformation of earthly sound.

The expulsion of oracular spirits from the world directly enables this transformation because their communications, even their bodies, were thought to be implicit in the medium of sound. In animist philosophies and religions, the breath or spirit of the world was believed to be present in all stages of material refinement, from plants and minerals to man, and beyond him, in beings of another sort: "Demons and genii were 'personalized' world spirit, clothed in air."[52] For Milton, spirits like these, while they abided in the air, threatened the universal voice that runs through all Creation. It cannot be redeemed while it also conveys the false, misleading intimations of contrary spirits. In his ode, they appear as the tutelary gods, whose words echo through the "arched roof" of the sky, and who populate the "haunted spring, and dale" and "urns, and altars round" (175, 184, 192).

When struck from the places and objects that contained them, these demons, because they are personalized expressions of the universal breath, emanate either in acoustical or meteorological form. The sounds they make as the earth releases them are everywhere. "A voice of weeping" and "loud lament" resounds over the land as the spirits flee the Christ-child, and "[t]he parting genius is with sighing sent" (183, 186). Similarly, the demons enclosed in the earth and its holy edifices "moan with midnight plaint" and make a "drear and dying sound" as they leave their containers (191, 193). The distinction between the breath of the world and the spirits themselves, already blurred in these descriptions of their departing sounds, is even further confused when Milton depicts them as emerging from within the very walls of their temples: "And the chill marble seems to sweat, / While each peculiar power forgoes his wonted seat" (195–96). The imagery in these lines is rather complex. The image of sweating expresses metaphorically the literal activity of the spirits leaving the stone and recalls the purgative effects of this physiological process. Simultaneously, the allusion to sweat on the "chill marble" recalls the formation of liquid from the air on cool surfaces, in other words, the process of condensation. The doubling up of biological and meteorological figures implies that, in addition to being forced out of the marble, much like beads of sweat from a body, the powers are also drawn out of the surrounding air and deposited on the stone like drops of dew. The "sweatings of Marble," as the naturalist John Beale would call them later in the century, were a famous weather prognostic of rain.[53] The spirits' emergence from both the stone and the air may be explained by Milton's view that demons are present in all of the elements, which he expresses in lines 93–94 of *Il Penseroso,* citing the authority of Hermes Trismegistus and Plato.[54] By portraying the overpowered spirits of the *Nativity Ode* as local genii and ingredients of the elements themselves, the ode asserts that before the birth of Christ, the purity and acoustical capacity of the climate was utterly compromised because it contained the bodies of demons and carried their deluding voices. The dilution of the world spirit by these evil agents weakens the unifying power of Nature's voice.

The additional suggestion that personalized spirit may be passed from the earth either through secretion, much like sweat, or the meteorological process of condensation, reinforces the association Milton makes at the beginning of the poem between exhalation and the voice of Nature. The descriptions of Nature exhaling the instructions for making snow and, in the last section of the ode, complaining as it sweats out bad vapors, imply that, in a very real sense, the earth has a breathing, perspiring, and speaking body that expresses itself vocally with its weather-causing exhalations. Milton's suggestion that exhalation provides a passage out of the world (as it does for "[e]ach fettered ghost" [234]) captures something of Lucretius's atomist explanation of the rapid formation of storm clouds: "It is no wonder then if often within a short time tempest and darkness overhanging above cover up sea and land with storm-clouds so great, since from all quarters through all the passages of the ether, and as it were, through the breathing-channels of the great world around, there are comings-in and goings-out for the elements."[55]

Milton was to revisit the idea of exorcism as a meteorologically and physiologically productive process in the masque performed for the Earl of Bridgewater in 1634 and later known as *Comus.* Having been sharply rebuked with "words set off by som superior power," the masque's villain fearfully observes that

> a cold shuddring dew
> Dips me all o're, as when the wrath of *Jove*
> Speaks thunder, and the chains of *Erebus*
> To som of *Saturns* crew.[56]

As in the *Nativity Ode,* the suffusion of the air with divinely powerful sounds initiates a process of disintegrating evil from the world that is simultaneously bodily and meteorological—Comus seems to sweat but calls his perspiration "dew." The analogy he draws between himself and "som of *Saturns* crew" continues the idea that divine or righteous speech actualizes itself and eradicates evil meteorologically: Jove "speaks thunder" and vanquishes the Titans.

Its ancient association with exhalation seems to license the idea for Milton that speech or sound of a certain virtue can dislodge evil from latent or incorporate states.

Though it is devoid of human speakers (apart from the "simply chatting" shepherds) the *Nativity Ode* is incontestably a sonorant work, abounding with representations of speech, music, echoes, screams and laments, even deafening blasts. These sounds and utterances emanate from giant figures—God, Fate, Nature, Apollo, to name a few—and intervene in cosmic processes: the setting of the stars, the rising of the sun, the falling of snow, the quaking of the earth, and the subduing of wind and sea. One might argue that the vocalization of these processes has the effect of domesticating the cosmos, or habituating the divine to the mundane world. Yet the inverse effect is more powerfully achieved. By endowing the universe with vocality and sonic might, Milton environmentalizes speech and sound. Acoustics are dissociated from the human subject and associated, rather, with the spaces they fill up. Milton's sound-centered poetics are architectural as well as environmental. The ode repeatedly depicts the reverberations of sound inside domes ("[b]eneath the hollow round / Of Cynthia's seat," or "through the arched roof" of the world) or with respect to other structures (on hearths, in altars and temples, and even in hell's "dolorous mansions") (102, 140, 175, 190, 192, 198). By defining sound within its context, Milton dispels the assumption that the air and its acoustics can be completely controlled and confined by man. There are flaws and virtues implicit in sound itself, which it absorbs from and transmits to its environment.

In underlining the acoustical dimensions of the environment, Milton stresses the basic dependency of voices on the air. In order to speak humans must draw in breath, and though we normally are not aware of it, our breath partakes of the greater respirations of the world. The *Nativity Ode* demonstrates that even Nature, who may seem our truest guide on earth and who indeed ministers the breath of life, is herself corrupted and weakened by adverse voices. On the other hand Christ, the very embodiment of the Logos, realizes simply by being born a state of perfect harmony in the world, thereby eradicating its desecrating sounds. But he leaves the archenemy of humanity intact, though symbolically confined underground, until he returns at the "world's last session." From this course of events we should not conclude that the climate has forever been healed. The subsistence of evil in Earth's soil indicates that after Christ, bad spiritual influence still permeates the air. Early modern science assumed that the ground is laced with meteorologically potent exhalations and minerals, which, as I discuss in chapters 3 and 6, are associated in Milton's poetry with atmospheric corruption and satanic activity in the world.

The captivation of the air, as we learn in the *Nativity Ode,* through compelling speech or exquisite music, is a critical strategy for shaping its moral and physical status and influence on others. As a fervidly devout and ambitious poet, Milton is interested in achieving and refining this sense of control. Without abandoning the environmental conception of sound developed in the *Nativity Ode* and some of his other university works, Milton goes on in *Comus* to investigate the experience of the human subject within the fallen climate and the capacity of the body's senses to navigate its compromised acoustics. Chapter 2 balances these concerns against some of the major the ideas and events of the Renaissance that gave rise to new ways of thinking about the air and its sounds.

CHAPTER 2

Early Acoustic Theory and the Aural Soul in *Comus*

Much of the critical commentary on Milton's 1634 masque, widely known as *Comus,* has focused on its "reformation" of the genre.[1] Invoking the pomp and spectacle typical of such entertainments only to reject luxury as part of the "antimasque," *Comus* subverts the genre's royalist conventions and glorifies the reformist virtues of temperance and chastity. In emphasizing the political topicality of *Comus,* which was originally performed for the first Earl of Bridgewater's installation as the president of the Council of Wales, scholars have demonstrated significant thematic parallels to cultural debates and scandal in the family of Milton's patron.[2] Limiting the masque's outlook to a generic or political statement, however, underestimates the degree to which even this "occasional" work addresses broader concerns that occupy the poet throughout his career. Assuming that *Comus* is "in some ways a very 'personal' document" to Milton with an intellectual milieu reaching far beyond the Council of Wales to the evolving discourse on the phenomenon of sound, this chapter proposes that a central focus of the masque is the problem of hearing and speaking reliably in a climate pervaded by spiritual evil.[3]

The association of perceptual unreliability with environmental corruption in *Comus* picks up an implicit premise of Milton's earlier poetry that subsequent works develop in line with scripture and natural philosophy: spiritual regeneration and degeneration on earth produce changes in the atmosphere and accompanying changes in its acoustics. By the time he wrote the masque for the Ludlow audience, Milton had already depicted feats of intellectual and moral heroism as extraordinary auditory or acoustical events with wide cosmological implications.[4] In the previous chapter we saw that the *Nativity Ode* depicts Christ's birth as eliminating the power of the ancient oracles by winning over the atmosphere with the exquisite sound of the angelic choir. Portraying the air as the guilty medium of unreliable prophecy, the ode also suggests that even after the Incarnation the atmosphere will remain susceptible to acoustical contamination, at least until the final redemption.

The anticipated future purification of the air, however reassuring in the abstract, leaves those who are presently cloaked in noise and atmospheric haze in a quandary. Milton briefly addresses this conundrum in "At a Solemn Music" by prescribing an active role for Christians as they wait for musical and eschatological resolution: answer God "with undiscording voice" (17). Voiced verse cancels the dissonance of the atmosphere through a process of verbal arrangement and harmonization. Yet as Milton's epic invocations concede, even poets need divine inspiration to make music from the fallen material of air.[5] In a world where divine guidance is not always obvious and corruptive forces compromise the atmosphere, how can individuals trust their own senses? This perennially Miltonic question receives robust consideration in *Comus*.

The fact that the Earl of Bridgewater's children acted and sang their own parts in the masque underscores the question of the hazards of self-expression. The decision broke with the norm of allowing aristocrats to dance but never to speak in public masques.[6] However, far from embarrassing the family, the children's roles have been said to contribute to the "sober religious realism" of *Comus*.[7] The Ludlow masque had a concrete and particular relevance to its first audience. Lady Alice, John, and Thomas Egerton, the three youngest children of the new president of the Council in the Marches of Wales retain their names and identities on stage. The plot device of their journey through the woods, though itself an invention, is occasioned by a real event, their father's assuming his political seat, and the narrative concludes with the children joining their parents at Ludlow Castle, their new home and the place where the masque was performed. These quasi-biographical elements create the sense that the characters' trial is both grave and real. But imbedded in the frame story is the fantastical tale of the Lady's detainment in the forest by a formidable sorcerer. The apparent contrast between the historicity of the protagonists' roles and the magic of the monsters in the wood would seem to cast the central trial and the dangers it involves as strictly symbolic. However, the Lady's confrontations with the deceptive and threatening environment of the Forest of Dean depicted in *Comus*, which foists sensations and effects on her that seem highly improbable to us, illustrate roughly two contending modes of knowing the natural world in the early part of the seventeenth century. At various points in the masque the Lady and her brothers are invited to analyze experiences with the tools of Renaissance magic as well as the relatively novel experimental method.

At the time of *Comus*'s composition in 1634, neither of these approaches to understanding Nature had eclipsed the other. Though the church officially condemned it, magic was widely studied and practiced in the seventeenth century at all levels of society and in a variety of ways. The recuperation of Neoplatonist writings and ancient works such as the *Hermetica* undertaken during the Italian Renaissance by Pico della Mirandola and Marsilio Ficino, and Heinrich Cornelius Agrippa's synthesis of the magical aspects of these works into a coherent "occult philosophy," contributed to a culturally broad European fascination with magic in the sixteenth and seventeenth centuries.[8] Magic had an unstable disciplinary and social status during this period, but nevertheless was integral to bringing about the transformation of natural philosophy into an experimental science. At the same time, the Scholastic philosophy maintained a foothold in the seventeenth-century pedagogy of science. In the early part of the century, the English universities were strongholds for Aristotelian thought, and, although they discredited scholastic methods, many pioneers of the experimental science, such as Francis Bacon, Galileo Galilei, and Rene Descartes, were in part indebted to Aristotle's ideas.[9] But the thorough revamping of scientific inquiry was already underway in the century before these individuals largely were active. Revolutionary developments in astronomy such as the publication of Copernicus's *De Revolutionibus Orbium Coelestium* (*On the Revolutions of the Heavenly Spheres*) in 1543, Tycho Brahe's observations of celestial phenomena in the 1570s and '80s, Johannes Kepler's calculations of the planetary motions, and finally, the shocking discoveries of Galileo's telescope, published in *Siderius Nuncius* (*Starry Messenger*) in 1610, had the gradual effect of undermining Aristotelian physics, establishing mathematics as fundamental to natural philosophy, and promoting the status of the mathematician to that of a full-blown philosopher.[10] While these momentous findings were well known (Copernicus's argument for heliocentricity traveled to England as early as 1576 in Thomas Digges's summary and translation) the experimental mentality which guided them was not yet institutionally sponsored (as it would later be by the Royal Society) nor directed toward a central school of thought in the first half of the seventeenth century.[11] Daniel Garber argues that none of the radical thinkers from the early part of the seventeenth century, strictly speaking, adhered to what later contemporaries and now historiographers refer to as the mechanical philosophy, one of the main intellectual movements to emerge from this period of philosophical upheaval, but rather "saw themselves as elements in this hurly-burly of anti-Aristotelian philosophies, each fighting against one another to become the new direction for philosophy to take."[12]

It was during these years of philosophical free-for-all that Milton composed *Comus* and the masque was performed. The masque's first printing in 1637 predates many of the founding documents of the new physics, for instance, Galileo's *Discorsi e dimostrazioni matematiche intorno à due nuove scienze* (*Two New Sciences*) and Descartes' *Principia Philosophaie*, which appeared in 1638 and 1644. Perhaps because Milton's work was conceived before the victory of experimental science could be definitively declared, and at a further remove from Boyle's corpuscular philosophy, its engagement with contemporary perspectives in natural philosophy has not attracted particular notice.[13] Nevertheless, in probing the limits and potential dilation of the human senses, Milton's masque displays an interest in what were known as the "mixed mathematical sciences," an Aristotelian category of knowledge that encompassed subjects such as astronomy, optics, music, and mechanics that were considered subordinate derivations of natural philosophy, but became increasingly relevant in the sixteenth and seventeenth centuries with the new emphasis in science on practical demonstration.[14] The era's revaluation of these subjects and the simultaneous reshaping of traditional learning by the integration of techniques from the mechanical arts reflect the humanist "tendency to consider practical science superior to theoretical knowledge."[15]

The reappraisal of practical or artisanal knowledge is particularly evident in the pathways along which the study of music, one of the "mixed sciences," evolved in the sixteenth and seventeenth centuries. During the Renaissance, the musician's concrete knowledge of their craft—formerly of no concern to the student of music theory—informed a new conception of music "in which sound is thought of in terms of movement" rather than number or ratios.[16] For centuries, the Pythagorean idea of harmony with its strictly numerical understanding of musically acceptable or "consonant" intervals held sway in music theory, but innovations in musical composition and instrumentation during the Renaissance demanded the reevaluation of musical sound on physical grounds.[17] The pioneers of this new science of music enlisted actual instruments in their tests, and rather than accepting that certain intervals are more pleasing than others because of their cosmological significance, strove to arrive at a theory of consonance by examining the vibrational behavior of strings.[18] Vincenzo Galilei (1520–1591), musician and father of Galileo Galilei, epitomized this new pragmatic impulse in refusing to accept the Pythagoreans' doctrine that only ratios of certain whole numbers produce consonant intervals and challenging this assumption through experiment.[19]

Before these methods yielded a mathematically and physically verifiable description of pitched sound—a task that would fall to the French friar and enthusiastic experimentalist Marin Mersenne—a variety of other models for conceptualizing sound were in play. A key alternative tradition was natural magic, which one scholar calls a "pre-modern form of natural science."[20] This specific kind of "white" magic, espoused by thinkers such as Giambattista della Porta, Agrippa, Robert Fludd, and to some extent, Francis Bacon, focused on harnessing the "occult" or hidden properties of nature to produce remarkable effects, and was distinguished from demonic forms of magic, which depend on invoking demons or making a pact with the devil.[21] Agrippa, who was widely read in England, offers a representative definition:

> Therefore natural magic is that which having contemplated the virtues of all natural and celestial things and carefully studied their order proceeds to make known the hidden and secret powers of nature in such a way that inferior and superior things are joined by an interchanging application of each to each; thus incredible miracles are often accomplished not so much by art as by nature to whom this art is as a servant when working at these things. For this reason magicians are like careful explorers of nature only directing what nature has formerly prepared, uniting actives to passives and often succeeding in anticipating results so that these things are popularly held to be miracles when they are really no more than anticipations of natural operations.[22]

Magicians of this stamp should be well versed in mathematics and physics as Agrippa points out elsewhere, "and knowing the middle sciences of both these, Arithmetick, Musick, Geometry, Opticks, Astronomie, and such sciences that are of weights, measures, proportions, articles and joynts, knowing also the Mechanicall Arts resulting from these, may without any wonder . . . do many wonderfull things."[23] Magic's association with mathematics and mechanics, indeed its frequent reliance on machines and artificial devices to bring out and display the secret powers implicit in nature, made it an important precursor to experimental science. In the words of John Henry, "the scientific worldview developed, at least in part, out of a wedding of natural philosophy with the pragmatic and empirical tradition of natural magic."[24] Penelope Gouk makes a related claim with respect to the scientific study of sound. She argues that the discipline of acoustics as it was first practiced in the seventeenth century was largely adapted from the tradition of natural magic.[25] Her research on the history of acoustics in England shows that the emergence of this science cannot be fully explained by the revolution in physics but depends on an inheritance of customs, demonstrations, and theory from the tradition of Renaissance natural magic, which valued the explanatory power of music and implemented musical instruments in its magical performances.[26] Thus, while individuals like Vincenzo Galilei and his son laid the theoretical groundwork for developments in modern acoustics, like Mersenne's laws of vibrating strings (mathematical formulae determining the potential frequency of a stretched string), a number of authors in a disparate magical tradition, including John Dee and Robert Fludd, developed a parallel idea of music that would also greatly influence "scientific" acoustical inquiry in the seventeenth century.[27] Works in this tradition uphold the mathematical and Neoplatonic concepts of music as a dynamic medium between macro- and microcosms and document the proliferation of musical instruments in the early seventeenth century by depicting their variety of forms and operations.[28]

My analysis of aural phenomena in *Comus* reveals Milton's familiarity with a variety of pre-Galilean acoustical theories, particularly Ficinian music-spirit theory and Francis Bacon's notes toward an acoustical program in *Sylva Sylvarum* (1627). By bringing out the masque's engagements with this eclectic body of literature, pieces of which get taken up by physicists in the late 1630s, this chapter argues that *Comus* reflects one of the main conflicts inherent in the science of sound at this transitional stage of its history—namely, the development of a practical science of music and acoustics in the absence of a mechanical theory of sound transmission. While the study of music gradually merged with the mechanical arts in the Renaissance, bearing out the humanist impulse to understand the psychological effects of sound and discover how best to control and use it, the coherence of acoustics as an objective science at the beginning of the seventeenth century was limited by the fact that its explanatory paradigms, even its experimental method, depended largely on magical and Scholastic concepts. The occult underpinnings of acoustic explanation have complex ramifications for a Puritan poet like Milton. He seems to have implicitly supported the philosophical aims of Bacon's new program of acoustical inquiry insofar as it employed rational means to increase the individual's powers of discernment through hearing and command over the voice. But the masque's occasional skepticism of the accuracy of sounds and the psychological consequences of music suggests that he also found cause for concern in the occult foundations of contemporary acoustical theory, and possibly, in its attitude toward the artificial manufacture of sound.

Francis Bacon is a pivotal figure in the murky, primal landscape of experimental acoustics. The subjects of music and sound occupy substantial space in his comprehensive work of natural history, *Sylva Sylvarum*, and portions of the *New Atlantis*, both published posthumously. Bacon's consideration of acoustics was not only extensive, but also revolutionary in its approach. His writings, for one, "were the first (in English natural philosophy) to discuss music outside the context of the mathematical ratio."[29] He rejects numerical, Pythagorean-

Platonic explanations of musical harmony because of their detachment from observable fact. In contrast to existing music theory, whose "mystical subtleties" fail to account for the actual experience of musical practice, his examination of music, he asserts, will "join the contemplative and active part together."[30] Accordingly, he proposes a vast number of acoustical experiments in *Sylva Sylvarum*, many involving instruments, others the human voice and various kinds of sound-carrying media, to investigate the properties of sound directly, and to determine how it might be altered or manipulated to serve human needs. The role that instruments and artificial techniques play in Bacon's program of acoustical inquiry recalls the emphasis on mechanical contrivance in musical writings from the magical tradition. Indeed, a large number of Bacon's experiments were taken directly from Giambattista della Porta's *Magia Naturalis* (1589).[31]

The concept of spiritus underlying Bacon's theory of sound transmission and perception grows out of a vitalist tradition that differs sharply from the mechanical philosophy. His explanations of the phenomenon known as musical sympathy (i.e., the sympathetic vibration of strings) and of hearing depend on the notion that all matter, whether animate or inanimate, contains pneumatic spiritus that has the ability to perceive stimuli.[32] The interaction of the "species" of sound, "an immaterial entity containing all the qualities of the sounding body within itself" with the spiritus found in the body and the external world is what accounts for auditory experience, according to Bacon.[33] Thus, the "species" of sound emitted by an instrument may cause a nearby instrument to resonate sympathetically by mingling with is spiritual, pneumatic parts; similarly with the human body, the species of sound enters the ear where it impresses itself on the body's spiritus, which in turn carries the sensation to the brain initiating the body's physical response.[34] Bacon subscribed to the magical views that spiritus was susceptible to astrological influences and might be acted on and altered by another person's imagination (or that of an intermediary).[35] The susceptibility of the spiritus in the human body to outward influences is an assumption of Bacon's acoustic theory. He follows Ficino in identifying musical harmony as the strongest stimulus to the spiritus, relative to, for example, taste or smell, reasoning that "the sense of hearing striketh the spirits more immediately than the other senses, and more incorporeally than the smelling."[36] More than just a sensation, music causes man to feel passions, because its impact on the spiritus is not momentary like that of an odor, but lasting: "harmony, entering easily, and mingling not at all, and coming with a manifest motion, doth by custom of often affecting the spirits and putting them into one kind of posture, alter not a little the nature of the spirits, even when the object is removed."[37] It is difficult to judge the strength of such beliefs, which were founded on principles from occult philosophy, given Bacon's occasional denouncements of magic and his larger vision for a New Science that privileges keen observation, logical method, and collaborative learning over knowledge acquired on authority or through auricular traditions.[38] Yet the "presence of both mechanist and dynamic vitalist conceptions of reality in Bacon's physics" is undeniable.[39]

Because he regards sound and music as subjects of natural philosophy, not mere illustrations of the mystical power of number, and anticipates many experimental techniques that would be used in subsequent years to forge a fully mechanistic account of sound, Bacon's acoustical program can be characterized as pioneering. But it also includes much lore from the tradition of natural magic and explains the operation of sound and hearing with concepts from Scholasticism and vitalism.[40] The coexistence of mechanistic and vitalist influences in Bacon's explanations of aural phenomena is also a feature of Milton's depiction of auditory experience in *Comus*. Bacon's unique treatment of acoustics represents the most current research on sound in publication at the time. Thus it provides a crucial comparison point to acoustical representations in Milton's masque.

Like Bacon, Milton is not content in *Comus* to examine music from a Pythagorean-Platonic perspective by presenting it purely as a means of elevating the mind and revealing the mathematical nature of the universe.[41] The exaltation of the intellect may be a desired consequence of song, but in practice, music is a product of sound. The masque measures music by its sensible properties: its ability to affect the passions, to travel long distances, to spring from refreshed spirits, etc. It also represents various other aural effects—the sound of riot, the hollering voice, echo, and silence—as objects of philosophical scrutiny, giving them a concrete existence outside of the perceiver's mind. Bacon calls the manifestation of sound outside the mind its "species" (as discussed above). The audible species is a quality distinct from the medium of sound that "behaves as though it is corporeal despite its immaterial nature" and acts on the perceiving body's spiritus independently of the air.[42] Bacon's belief that sound is not a motion, but a thing or quality that exists before it enters the perceiver's mind was characteristic of early theories of acoustics and distinguishes them from the mechanical explanations of Mersenne, Galileo, and Descartes, who viewed sound as the mind's translation of a simple movement of the air.[43] These two positions, however, were not the only viewpoints on the physical nature of sound in the sixteenth and seventeenth centuries.

Ficino's music-spirit theory represents a conceptual middle ground between an acoustical model like Bacon's, in which the species of sound but not the moving air mingles with the perceiver's spirit, and that of the mechanical philosophers who contended that the hearer only perceives percussions of air but that nothing material or immaterial actually enters the ear. Ficino contends that both air and movement are involved in the process of hearing. Both factors are crucial to his explanation of why music, more than anything else admitted by the senses, affects the spirit and through it, the soul. The similarity between the substance of music, "warm air, even breathing," and the airy substance of the spiritus, ensures its direct conveyance into the listener's spirits and transmission through the whole body.[44] Additionally, the fact that music "transmits movement and is itself moving" enables it to mimic the motions of the soul and spirit.[45] Ficino's discussion of the uniquely affective quality of music in his commentary on Plato's *Timaeus* demonstrates that the movement and aerial nature of song (along with emotional and intellectual content) are key sources of its power over the soul:

> But musical sound by the movement of the air moves the body: by purified air it excites the aerial spirit which is the bond of body and soul: by emotion it affects the senses and at the same time the soul: by meaning it works on the mind: finally, by the very movement of the subtle air it penetrates strongly: by its contemperation it flows smoothly: by the conformity of its quality it floods us with a wonderful pleasure: by its nature, both spiritual and material, it at once seizes, and claims as its own, man in his entirety.[46]

Ficino's belief that musical sound is made of refined air is at once a forward- and backward-looking assertion as it draws on a non-Aristotelian strand of acoustical theory dating back to the Middle Ages that regarded sound as subtle or fine air and simultaneously anticipates mechanistic accounts of sound by recognizing that it depends centrally on the movement of air.[47] Sound in *Comus* often exhibits the spiritual and material qualities Ficino gives it: likeness to air, likeness to spirit, and motility. Furthermore, elements of the masque that suggest music may be used for ill, to deceive, impair, or entice the listener, may be traced to Ficino's notion of musical sound as almost capable of undoing the soul by infusing the spirits with its own movements.

"WITH PUISSANT WORDS, AND MURMURS": THE POWER OF INCANTATION

By framing the Lady's trial not only as a test of the will, but also as a challenge to her aural faculties (her ability to sing, trust her ear, speak out in her own defense, and then again, keep silent), Milton takes a Baconian view on the acoustical value of music, foregrounding the role that sound and experiments involving sound play in determining one's physical and moral position in the world. Understanding aural experience from a physical standpoint, as Bacon and others strove to do in the seventeenth century, is an aspiration that Milton shared. But even as it registers approval of this kind of experimental inquiry, Milton's masque is deeply guarded about the theory informing this early research and the potential misuses of acoustical artifice. These reservations grow out of the association of occultism with the history of acoustical explanation and musical experiment.

I do not wish to paint Milton as a skeptic who regarded magic as powerless to alter or manufacture sounds and affect the hearing of individuals. Nowhere in Milton's writings does he explicitly deny the efficacy of magic. His interpretation in *Areopagitica* (1644) of Acts 19:19, which describes Christian converts from Ephesus burning their magical books, balks at the idea of the state censoring such books, but nevertheless indicates that the practice of magic is a truly dangerous affair: "As for the burning of those Ephesian books by St. Paul's converts, 'tis replied the books were magic—the Syriac so renders them. It was a private act, a voluntary act, and leaves us to a voluntary imitation: the men in remorse burnt those books which were their own; the magistrate by this example is not appointed; these men practiced the books; another might perhaps have read them in some sort usefully."[48] Milton could have argued that the magistrate's interference is unnecessary in the case of magical books because they are harmless or ineffectual. But he makes no such comment. Far from being vacant tomes, the books have a capacity to incite magical practice that individuals may privately seek to eliminate by burning them; moreover, if the books are read without involving the reader in idolatry or commerce with evil spirits, they may even be found to contain useful knowledge.

The distinction Milton draws in *Areopagitica* between books used for magical practice and those mined for knowledge also applies in *Comus* where the enchanter's spells are clearly regarded as illegitimate and deplorable examples of necromancy, but the shepherd lad's knowledge of a thousand simples represents an acceptable and valuable appropriation of magical lore. If we map the shepherd boy's pharmacopeial knowledge and Comus's magical art onto divisions in occult philosophy, the former falls safely in the realm of natural magic, the latter in the category of black magic or sorcery.[49] These types of magic recognize different sources of power: the natural magician locates it in the hidden virtues or occult properties of Nature and the necromancer in the power of the devil or evil spirits.[50] The shepherd lad's expertise clearly pertains to the "strange and vigorous faculties" inherent in Nature and enables him to pluck out from his collection a root "of divine effect" for Thyrsis (628, 630).

Sound and music were key subjects of natural magic, a quasi-scientific body of thought whose intellectual aims were not so different from those expressed in Milton's program of "universal learning" from *Prolusion VII*. The following is a definition of magic given by the natural magician Porta: "I think that Magick is nothing else but the survey of the whole course of Nature. For, whilst we consider the Heavens, the Stars, the Elements, how they are moved, and how they are changed, by this means we find out the hidden secrecies of living creatures, of plants, of metals, and of their generation and corruption; so that this whole Science seems meerly to depend upon the view of Nature, as afterwards we shall see more at large."[51] But some authors in the occult tradition were interested in sound as a way to facilitate ceremonial magic. This type of magic was more questionable than natural magic as it used incantations and other rites to invoke the aid of spirits or other intelligent influences. Ceremonial or spiritual magic supposedly allowed witches or black magicians to commune with bad daemons.

But it might also be employed to procure the influence of good spirits. The third section of Ficino's *De triplici vita* (*Three Books on Life*) (1489), titled *De vita coelitùs comparanda*, exemplifies the latter kind of ceremonial magic because it proposes various "ways of attracting benevolent celestial influences—chiefly but not exclusively planetary—and repelling maleficent ones."[52] In the *De vita coelitùs comparanda*, Ficino instructs the reader how to perfect the harmony between one's life and the stars' ruling influence by discovering which songs or tones belong to specific stars so they may be addressed to the right parts of the sky.[53] Agrippa affirms that words arranged in verses that properly conciliate the stars and their intelligences have a unique power to enchant: "Such like verses being aptly, and duly made according to the rule of the Stars, and being full of signification, & meaning, and opportunely pronounced with vehement affection, . . . and by the violence of imagination, do confer a very great power in the inchanter, and sometimes transfer it on the thing inchanted, to bind, and direct it to the same purpose for which the affections, and speeches of the inchanter are intended."[54] He locates this power—following Ficino's description of music very closely—in the living, moving, spirit-like quality of the words, which have been infused with both human reason and celestial virtue:

> Now the instrument of the inchanters is a most pure harmoniacall spirit, warm, breathing, living, bringing with it motion, affection and signification, composed of its parts, endued with sence, and conceived by reason. By the

quality therefore of this spirit, and by the Celestiall similitude thereof, . . . verses also from the opportunity of time, receive from above most excellent vertues, and indeed more sublime, and efficacious then spirits, & vapors exhaling out of the Vegetable life, out of hearbs, roots, gums, aromaticall things, and fumes, and such like. And therefore Magicians inchanting things, are wont to blow, and breath up on them the words of the verse, or to breath in the virtue with the spirit, that so the whole virtue of the soul be directed to the thing inchanted, being disposed for the receiving the said vertue.[55]

Voiced sound, by Agrippa's account, amounts to far more than the mechanist idea of a motion of the air sensed by the ear and translated in the mind. It is comparable to vapors from plants and to perfumes, but consists of a higher, astral substance. When the words of an enchantment are pronounced, breathing in and blowing out transfers and infixes their virtue. Agrippa's alarming suggestion that words, because of their airy substance, may be mixed with celestial influence to work involuntary effects on the spirit, haunts Milton's poetry. By having the Lady and her brothers confront the son of Circe, "[d]eep skill'd in all his mothers witcheries," in a dark forest where their powers of vision are greatly diminished Milton deliberately tackles this fear head on (512). The children's quandary raises questions about sound that interested contemporaries but by no means were satisfactorily understood, certainly from the perspective of early modern science, but also from Milton's standpoint, that of moral philosophy and poetry. How reliably does sound inform the judgment? How strongly does it move the passions and through them the soul? How does one speak virtuously in a spiritually compromised and, thus, acoustically fragile environment?

THE HAZARDOUS ACOUSTICS OF "SPUNGY AYR"

From the very beginning of *Comus*, the possibility that sound on earth may not be communicated directly or faithfully from its source is raised. The problem lies with the medium of sound—the air. In the first lines of the masque, the Attendant Spirit complains about bad air quality, contrasting the "[r]egions milde of calm and serene Ayr" from which he originated with Earth's "smoak and stirr" (4–5). His "bright aëreal" nature is indisposed to "the rank vapours of this Sin-worn mould"—the fumes that course through the mortal body and the material world—but he suffers them in order to lend his special aid to a just few on earth (3, 17). The Lady echoes these sentiments when she opens her mouth to answer Comus's argument about the necessity of enjoying Nature's riches. She confesses that she "had not thought to have unlockt my lips / In this unhallow'd air," until realizing that virtue needs a champion (755–56).

If the Attendant Spirit and the Lady characterize the air and the vapors entering and exiting the body as putrid and adverse to health, then the picture that emerges of the wild wood at evening is of a darkened, delusive atmosphere inimical to clear perception. The faculty of sight is practically useless in this setting. The "blind mazes" of the wood are inscrutable to the eye, and the night is so dark that the Lady exclaims,

> O theevish Night
> Why shouldst thou, but for som fellonious end,
> In thy dark lantern thus close up the Stars,
> That nature hung in Heav'n.
> (194–97)

Her remark is echoed by the older of her two brothers, whose prayer for starlight points up the thickness of the forest canopy and the pitchy blackness of the night:

> Unmuffle ye faintstars, and thou fair Moon
> That wontst to love the travailers benizon,
> Stoop thy pale visage through an amber cloud,
> And disinherit *Chaos*, that raigns here
> In double night of darknes, and of shades;
> (330–34)

Lack of light impels the wanderers to lean more heavily on their sense of hearing for guidance. The Lady expresses this inevitability as she tries to locate those who are responsible for making the sounds of merriment she hears issuing from the wood: "This way the noise was, if mine ear be true, / *My best guide now*" (169–70; emphasis added). The Attendant Spirit in the person of Thyrsis also uses his ear to locate the Lady:

> Then down the Lawns I ran with headlong hast
> Through paths, and turnings oft'n trod by day,
> Till *guided by mine ear* I found the place.
> (567–69; emphasis added)

Eyes have been rendered useless by darkness.

By depicting the wood that separates the children as stubbornly tangled and wrapped in impenetrable darkness, the masque raises the possibility that sound and hearing will help to unite them. In response to his older sibling's improbable suggestion that they might see

> som gentle taper
> Though a rush Candle from the wicker hole
> Of som clay habitation,
>
> (336–38)

the younger brother expresses doubt about gaining visibility in the wood, hoping, rather, that sound might reach their ears:

> Or if our eyes
> Be barr'd that happiness, might we but hear
> The folded flocks pen'd in their watled cotes,
> Or sound of pastoral reed with oaten stops,
> Or whistle from the Lodge, or village cock
> Count the night watches to his feathery Dames,
> T'would be som solace yet, som little chearing
> In this close dungeon of innumerous bowes.
>
> (342–48)

All of the opportunities he imagines for aural perception are plausible—from the noises of farm animals to the sounds of the shepherds and villagers. And indeed, it is Thyrsis's far away "hallow" that finally rescues them from complete darkness, literal and metaphorical (480).

But even if aural phenomena seem to offer consolation to the children and a way to find their bearings in the dark woods, the masque shatters these assumptions by illustrating how sound can mislead and even infect the mind. Comus's opening song and the aural and visual illusions he presents to the Lady demonstrate that sound cannot replace light in its absence; nor can it alleviate entirely the sense of misdirection or "*Chaos, that raigns*" at night. Certain sounds are actually distorted by the night air, others suspended and preserved by it. Darkness and thick air present a biding place for sound where it lingers and sometimes transmutes even after its source has been removed.

The synesthetic language Comus uses to describe how the shadiness of the wood conceals his followers' illicit acts introduces the idea that darkness may serve to trap or retain sound, whereas light readily releases it. When he bids his fellow revelers to begin their sacred rites, he reassures them that "[t]is onely day-light that makes Sin / Which these dun shades will ne're report" (126–27). He continues to conflate speech with illumination when he promises the goddess Cotytto that they will complete all her rites

> Ere the blabbing Eastern scout,
> The nice Morn on th'*Indian* steep
> From her cabin'd loop hole peep,
> And to the tel-tale Sun discry
> Our conceal'd Solemnity.
>   (138–42)

When the Lady arrives at the place where she thought she heard riotous sounds, darkness seems to act as Comus suggests, mysteriously absorbing any trace of the sounds' origins:

> This is the place, as well as I may guess,
> Whence eev'n now the tumult of loud Mirth
> Was rife, and perfet in my list'ning ear,
> Yet nought but single darkness do I find.
>   (200–203)

But the Lady's words "eev'n now," which indicate that the ruckus she heard is *still* perceptible, present a point of confusion since they are succeeded by the words "[w]as rife," which suggest conversely that the sound is no longer distinct. Is she still hearing the sounds that Comus hushed up just before she arrived? Possibly. The phenomenon of sound delay had been discussed by Bacon who knew that sound traveled slower than light and even proposed a way to measure its speed.[56] In *Sylva Sylvarum*, he observes that the "species of audibles do hang longer in the air than those of visibles."[57] The persistence of sound after it is created may relate to the thickness of the medium and the time of day. He explains, "in the day, when the air is more thin, no doubt, the sound pierceth better; but when the air is more thick, as in the night, the sound spendeth and spreadeth abroad less"; "thick air preserveth the sound better from waste."[58] These accounts of the persistent nature of sound, particularly at night, help us to see why the clamor of Comus's party is still audible to the Lady even when the partiers have vanished and ceased to make noise. Already inclined to lag and linger in the air, the species of sound is further preserved from dissipation by the insulating material of darkness. That the Lady switches to past tense ("the tumult . . . / Was rife") as she narrates her experience, however, suggests that either the sound eventually falters or the Lady loses confidence in what she hears. The behavior of aural stimuli—the fact that they hang around in the air even when their producers no longer remain—imposes a disruption between sound and sight, possibly causing the Lady to doubt her ear, which she previously presumed was her "best guide."

Milton is not content, however, to admonish us of the disorienting physical properties of sound as delineated by Bacon. He seems to discern even greater danger in certain characteristics of sound that Bacon is happy to describe from a physical standpoint without discussing their moral or spiritual implications. Throughout his natural philosophy, Bacon is at pains to distinguish his method from a certain fraudulent sort of natural magic that he believes acts on men's critical faculties like a drug because it rests on opinions and mystical truths rather than the active and rational pursuit of real knowledge.[59] Perhaps because of his contempt for this delusive kind of magic, which he views as an aberration from the original conception of magic, Bacon avoids speculation about the interaction of sounds with astral or spiritual influences, which, as we have already seen, interested occult philosophers. In *Comus*, Milton goes beyond Bacon's contentions that sound is not communicated instantly, that it lingers in the air, and that dark and dense air is effective for retaining sound, in implying that the substance of sound implicit in the air is susceptible to spiritual influence and manipulation—particularly by magical means. In recognizing the potential for spiritual agents or meddling magicians to interfere with sound, Milton, who in many other respects accepts Bacon's acoustics, advances arguably a less naïve model of sound by revealing how even a physical account of aural phenomena like Bacon's is compatible with the operation of magical and supernatural forces. Bacon fails to acknowledge that by giving sound a life outside the ear in the mysterious immaterial vehicle of the species, and by allowing this entity to pass into the body and work on the spirit, he invites the notion that the species—and through it, the human soul—may be altered or corrupted by spiritual means.

Milton, on the contrary, is willing to entertain the notion that aural phenomena, while explicable through natural laws, are also subject to magic, which as a science furnishes useful ways of explaining sound, and as an art is capable of affecting and diverting sounds to serve the magician's purposes. That magic is afforded this power in the masque, particularly the power to draw out or exploit certain properties inherent in the environment, is evident from Comus's ability to channel corrupted sounds into the Lady's mind. This occurs just as the Lady seems to indicate that the sound, whose producers she wandered in search of, has subsided. At this point the invisible residue of that commotion "[throngs] into [her] memory" confronting her with "calling shapes, and beckning shadows dire, / And airy tongues, that syllable mens names" (205–7). These phantom voices are not merely reverberations of the sounds the Lady heard earlier; they have been transformed into phantasms that conduct themselves into the Lady's mind where they call and beckon to her as though of their own agency. How did the sound she originally heard transform from "[m]idnight shout, and revelry" into the sinister whispers of "airy tongues" (103)?

The seeds of the enchantment may be traced to Comus's invocation of Cotytto, "Goddesse of Nocturnal sport," which he plies just before the Lady's footsteps are perceived nearby (128). In addressing the deity, Comus stresses that this conjuration always occurs at the darkest point of night:

> mysterious Dame
> That ne're art call'd, but when the Dragon woom
> Of Stygian darkness spets her thickest gloom,
> And makes one blot of all the ayr.
>
> (130–33)

Comus's rather disgusting figure of a hell monster's womb spitting out a thick cloud of darkness is not just the macabre expression of a lascivious sorcerer; it displays his continual desire for the protection of darkness and anticipates his request for additional cloud cover. In the next line he begs Cotytto to

> Stay thy cloudy Ebon chair,
> Wherin thou rid'st with *Hecat'*, and befriend
> Us thy vow'd Priests.
>
> (134–36)

The elder brother's speculation that the stellar and lunar influence he seeks to light the way back to his sister may "be quite damm'd up / With black usurping mists" seems to be in response to this dark, obscuring fog, which Comus has invited to settle over the wood (335–36). The fact that it blocks the starlight from reaching the children is no accident. As the Lady rightly suspects, the night has eclipsed the stars "for som fellonious end." That is to say, the fog Comus summons from the Thracian deity and her associate Hecate, goddess of sorcery, is clearly imbued with all the physical hazards of a dark haze—including the capacity to amplify, prolong, and preserve sounds as Bacon suggests—but also, the threatening enchantments of black magic.

Comus uses other magical means to increase the already bewildering nature of nighttime haze. In his opening speech, which precedes a lively song, he brags with Cavalierish pomp, "[w]e that are of purer fire / Imitate the Starry Quire" (111–12). As Blair Hoxby argues, the rite Comus initiates here illustrates a key function of balletic dance in Stuart masques: the magic conferral of astral influence on the masquers via their figural imitation of stellar movements.[60]

Comus appears to subscribe to Ficino's view that through song we imitate and channel astral influence most effectively.[61] In addition to attempting with his antic measure to infuse his crew with celestial power, Comus further alters the atmosphere of the wood by seeding it with magical spells or charms. As he hears the Lady approaching, Comus tosses "dazling Spells into the spungy ayr / Of power to cheat the eye with blear illusion" and cover up the remnants of his crew's revelry (154–55). I argue that these spells, which Comus casts aloft with a certain "[m]agick dust," do more than interfere with the enchanter's outward appearance and that of the dark forest (165). That they are visually effective is evident from the success of Comus's shepherd disguise and the Lady's conviction that the "sable cloud" overhead "[turns] forth her silver lining on the night"—quite possibly the result of the magic Comus has performed on the air (222–23). But they also carry aural implications, as magical verses themselves, Agrippa says, are made of "a most pure harmoniacall spirit, warm, breathing, living, bringing with it motion, affection and signification." Moreover, Comus tells us that his spells have been absorbed into the "spungy ayr," the medium through which all the other sounds in the vicinity have moved or will move. Milton's word choice here is significant. By likening the dense atmosphere of the wood to a sponge, he emphasizes its power to absorb and retain Comus's spells and the sounds of his revelers. Whether these suspended sounds consist of species, as Bacon would have it, or refined and moving air, as Ficino believed, they are compromised by the spells that are also stuck in the enchanted fog. The astrologically empowered spells and "black usurping mists" that Comus uses to accentuate the natural acoustics of dark air are responsible for the moment when the revelers' tumult ceases to ring "eev'n now" in the Lady's ear and beckoning voices begin to flow into her imagination.

The air, for Comus, is a place of spiritual potential. It is a region full of animate spirit, celestial and daemonic, which the enchanter may constrain and divert for his purposes. The spirituous substance of sound is a crucial vehicle for inviting disembodied spirit into objects, persons, or locales, and similarly, for releasing it from them. Comus is well aware of the utility of sound as an instrument of enchantment and uses it to degrade and deplete his victims' spirits as well as harvest spirit from the environment. His belief that the air is saturated with spirit that is responsive to sound is evident not only from his tuneful invocation of the stars and the goddesses of the night, but also his curiously tactile descriptions of the silent night air. When he overhears the "raptures" that the Lady sings to Echo he remarks,

> How sweetly did they float upon the wings
> Of silence, through the empty-vaulted night
> At every fall smoothing the Raven doune
> Of darknes till it smil'd.
>
> (246, 248–51)

He says the night is vacuous or "empty-vaulted," but characterizes its negative qualities in material terms: silence has wings; darkness is downy. This contradiction recalls the paradoxical image in *Paradise Lost* likening hell to a furnace whose flames emit "[n]o light, but rather darkness visible" (1.63). Though a better term for the oblivion in Milton's masque might be "darkness palpable," the underlying meaning is similar. There is no light; nevertheless the darkness is crammed with perceptible activity. Here that active presence is registered by beating wings and raven feathers, which are stilled or coaxed into ordered movements by the caresses of the Lady's voice.

Comus alludes again to the feathered texture of the sky in his argument against virginity when he claims that, if Nature's sons were to refuse her gifts, the whole earth would be strangled with her fertility and "the wing'd air dark't with plumes" (729). On its face, this statement is a simple prediction that the earth's bird population would explode, but it also conjures an image of a dark pall of wings in the sky. That Comus professes to fear an atmospheric eclipse is ironic since he is the chief beneficiary of obscuring shade and darkness. Moreover, if the Elder Brother imagines correctly, the sorcerer should be accustomed to being encircled by winged creatures (along with other demons) as he commands armies of "*Harpyies* and *Hydra's*, or all the monstrous forms / Twixt Africa, and Inde" (604–5). By portraying the crowding of the sky with such forms as a calamitous outcome of self-denial, Comus seems to admit—however insincerely—that the benighted existence he leads in the wood enshrouded by tree boughs, magical fog, and the rustle of spiritual wings, is truly a kind of suffocating hell.

Comus displays awareness of his faults most apparently when he hears the Lady sing or speak. The divine quality of her voice strikes him as serving a dramatically different purpose from that for which he employs sound. On hearing her sing, he takes her for an exotic creature, or else,

> the Goddes that in rurall shrine
> Dwell'st here with *Pan*, or *Silvan*, by blest Song
> Forbidding every bleak unkindly Fog
> To touch the prosperous growth of this tall Wood.
>
> (266–69)

It seems that Comus does not have a particular goddess in mind; rather, amazed with the Lady's voice, he invents what he thinks is a worthy identity for her, one that reflects her powerful vocal capabilities. Notably, the genius he describes uses her "blest Song" to chase away unfriendly fog, the very meteorological phenomenon he has just invited Cotytto to cast over the woods. Comus's comment reflects his real concern that the Lady's divine voice will disperse the mists he has drawn as a curtain around his nest of sin. His intuition echoes a view held by some at the time that sacred music has the power to ward off evil spirits. In his *Principles of musik, in singing and setting* (1636), Charles Butler designates casting out the devil and calming the possessed as "[e]xtraordinari" uses of divine music, citing the curative effects of David's harp on the troubled Saul as well as Luther's belief that "[this] power of Musik against evil spirits . . . stil remain" to this day.[62] Milton's notion of music working on behalf of virtue, however, is more complex than Butler and Comus would have it. When the musical and verbal parts of a song reinforce each other—the words receiving virtue from the sound of music and music benefiting from the meaning of words—it can possess a universal appeal. Unlike a virtuous plant such as Haemony, which exhibits antipathy to some objects and sympathy to others, divinely inspired sound, like Orphic music, holds sway over all who hear it, bringing them to life and drawing them towards good. The universally attractive power of divine song explains why Comus—himself a depraved agent of evil—is simultaneously in awe of and allured by the sound of the Lady's music ("such Divine inchanting ravishment") (244).

If Comus and the Lady are both empowered by acoustics, but the Lady's words are "set off by som superior power," namely Virginity, why doesn't the masque culminate with the Lady triumphing over Comus with her verbal talents (800)? An oratorical victory would seem to illustrate the ineffectualness of Comus's "well plac't" and "[b]aited" words against the Lady's divinely guided sounds (161–62). But Milton stops short of proving that sound can be a more powerful instrument in the hands of a good agent rather than an evil one. We are asked to take this on faith when the Lady promises Comus that, should she choose to speak, her true cause

> would kindle [her] rap't spirits
> To such a flame of sacred vehemence,
> That dumb things would be mov'd to sympathize.
> (793–95)

Foregoing this outcome, however, the masque suggests that trouncing Comus with inspired sound might come at a personal cost to the Lady. Thus, let us consider how the masque explores alternative uses of sound that avoid such a fate while conserving and even enlarging the spirit.

"NOT A WASTE, OR NEEDLESS SOUND": THE AURAL ECONOMY OF SPIRIT

The Lady's final pronouncement, which literally causes Comus to break out in a cold sweat, illustrates the central role that spirits play for Milton in the transmission of sound. Comus's physiological response to her claim that the "brute Earth," will move sympathetically to the vibration of her "rap't spirits," completes a chain reaction initiated by the sound of the Lady's voice (793–802). This moment recalls the phenomenon of sympathetic resonance that Bacon and others before him observed in the responsive behavior of musical instruments.[63] As one lute string transfers its motion to a similarly tuned string on another instrument, the Lady imagines that her divinely attuned voice, through its harmony with the earth, could inspire it to shake and crumble over her assailant's "false head" (798). A mechanical explanation for the process of sympathetic resonance was unknown at the time *Comus* was written. But a theory of the spiritus like Bacon's could explain how sound causes responsive motion in nonliving, "dumb things." He believed that "inanimate bodies respond to sound and resonate or produce echoes because the species of sound mingles with the pneumatical part of the body, its *spiritus*."[64] Had Milton fully subscribed to this notion of resonance and used it to describe the operation of the Lady's voice, he might have deemed it safe for her to say her piece and "unfold the sage / And serious doctrine of Virginity" (786–87). While the objects receiving sound in Bacon's account suffer an involuntary response, the sound maker undergoes no unusual effects.[65] But Milton seems to think that the act of speaking also involves an outlay of spirits on the sound maker's part, the loss of which the Lady is unwilling to suffer for the pacification of a hopelessly depraved soul like Comus. Her allusion to the phenomenon of spiritual rapture, which can involve the departure of the spirit from the body, corresponds to other moments in the masque where sound is associated with spiritual dislocation. As we shall see, the possibility that aural experience may deplete the body of its spirits—along with the fear that they may be corrupted—is a major concern for speakers and listeners in Milton's masque, shaping their decisions about when to speak and listen attentively.

The first indication that bodily spirit is a limited resource in *Comus* appears in the Lady's statement that, though she does not have the strength to yell for her brothers, she will try to sing loudly to them because her "new enliv'nd spirits / Prompt [her]" (227–28). Prior to feeling emboldened, her spirits were endangered by Comus's enchantments, which succeeded in startling her thoughts, though they intended much worse—to "astound" or freeze her mind (210). At last her spirits are "new enliv'nd," saved from the brink of astonishment by her conscience, which re-minds her (quite literally) of her heavenly champions (Faith, Hope, and Chastity). The association of lively spirits with the ability to cast one's voice so it may "be heard farthest" demonstrates that bodily spirit is a wellspring of vocal force and implies that paucity of spirit, by the same token, may jeopardize the voice (226). In light of Ficino's highly influential account of musical sound as elementally similar to the spiritus, we may construe the Lady's voice as an emanation of her spirits. The idea that speaking volubly was a physically depleting activity was anticipated in classical philosophy. Lucretius contends, for example, that in speaking we actually give up a part of our selves: "it does not escape you how much body is taken away and drawn off from men's very sinews and strength by speech continued without pause from the glimmer of rising dawn to the shades of dark night, above all if it is poured out with loud shouting. And so the voice must needs be of bodily form, since one who speaks loses a part from his body."[66] Even graver than loss of body, though related to it, is spiritual exhaustion, the consequence in *Comus* of over-extending oneself in song or speech.

When the Attendant Spirit advises the Lady after she has been freed from her paralysis—"Not a waste, or needless sound / Till we com to holier ground"—he alludes to the fact that speaking could deprive her of spirit as well as the possibility that making sound in the haunted forest invites predation (941–42). Earlier, when he heard the Lady's song wafting above the wood, he was alarmed at how "neer the deadly snare" she sang (565). Yet detection is sometimes desirable. While the Lady's voice initially attracts Comus's notice, it also plays a role in helping the Attendant Spirit track her down: "guided by mine ear I found the place" (569). More worrisome than being found out is the possibility that sound has an evacuating effect on the body's spirits. Comus's recollection of his mother's relationship to music illustrates the danger sound can pose to the spirit by supplanting its functions and even its substance:

> I have oft heard
> My Mother *Circe* with the Sirens three,
> Amidst the flowry-kirtl'd *Naiades*
> Culling their Potent hearbs, and balefull drugs,
> Who as they sung, would take the prison'd soul,

And lap it in *Elysium*, *Scylla* wept,
And chid her barking waves into attention,
And fell *Charybdis* murmur'd soft applause:
Yet they in pleasing slumber lull'd the sense,
And in sweet madnes rob'd it of it self.
(251–60)

These Homeric singers do nothing to liberate the "prison'd soul," though their potion-like song surrounds it with heavenly sensation. Comus describes their bewitching music with images that are heavy and liquid in contrast to the aerial terms that the Attendant Spirit applies to the Lady's song, "a soft and solemn breathing sound / . . . like a steam of rich distill'd Perfumes" (554–55). Circe's music puts the sense to sleep, in the vein of the frivolous sort of natural magic that causes men's understanding to slumber according to Bacon. The parallels between Milton's account of the sorceresses' song and Bacon's indictment of false magic are quite striking. Here is the passage from Bacon's *Of the Dignity and Advancement of Learning* (with added italics):

> But this popular and degenerate natural magic has the same kind of effect on men as some soporific drugs, *which not only lull to sleep, but also during sleep instill gentle and pleasing dreams*. For first it *lays the understanding asleep by singing* of specific properties and hidden virtues, *sent as from heaven* and only to be learned from the whispers of tradition; which makes men *no longer alive and awake* from the pursuit and inquiry of real causes, but to rest content with these slothful and credulous opinions; and then *it insinuates innumerable fictions, pleasant to the mind*, and such as one would most desire,—like so many dreams.[67]

Interestingly, Bacon employs the concept of song in this analogy as one of the vehicles whereby magic lulls the mind into a complacent sleep, though he is careful to exclude from his discussion of acoustics in the *Sylva Sylvarum* explicit mention of the magical uses or properties of sound. In this passage, Bacon caricatures the actual instruments of magic—sleep-inducing drugs, music, whispered words, and manufactured dreams—stripping them of any mystical potency and portraying them as a smokescreen for the truly stupefying logic of certain magical traditions, which rest on received opinion. Milton certainly shares Bacon's scorn for the deception definitive of certain kinds of magic, but he does not deny that their instruments (music, drugs, and spells) can affect the spirits. In Comus's speech, Milton openly recognizes the kinship between magic and music, showing how in practice song can effectively execute the ends of sorcery. Music is not simply a metaphor for magic in *Comus*. It can serve as a real, subtle, and insidious means of performing magical enchantment.

How exactly does it entrance the listener? It is not just with drowsiness that Circean music overcomes the sensitive faculties. The sirens' singing also instills a "sweet madnes" that robs the sense "of it self." Madness, here, refers to the notion of ecstasy or rapture—the separation of the soul from the body initiated by a ravishing experience.[68] Music can trigger ecstasy, which was often construed as a means of gaining prophetic insight or closeness to God. Although it is not entirely clear what Milton means by "sense" in this passage, in the context of rapture we may presume that he is speaking about the lower, irrational soul—what we have been calling the spiritus—or, more specifically, the animal spirits of medical theory, which were thought to control sense perception and the imagination among other things.[69] According to a long tradition, developed by the Neoplatonists, dealing with the spiritual container of the soul and its potential to detach itself in ecstasy, man could have divinatory dreams and commune with spirits and demons when his *spiritus phantasticus* or imagination left his body in rapture.[70] In such a dream-state, "the soul has revelations . . . because it is no longer hindered by the corporeal senses—'a pilgrim from the flesh.' . . . [T]he silence of the external senses allows the imagination . . . to be moved by superior or divine agents."[71] This impairment of the corporeal senses along with the departure of the soul is likely the kind of spiritual displacement to which Comus alludes when he talks about his mother's music robbing the sense "of it self." Circe's song, whose power is strengthened by her magical drugs, reduces the listener's senses by extracting the sense-giving spirit from them in a "sweet" but irrational act of larceny. In order to glut themselves of the sound, the senses surrender the body's spirit and therefore their capacity to perceive. The "guilefull spells" Comus uses "[t]o inveigle and invite th'unwary sense" of travelers stranded in the forest also employ Circe's method, "invit[ing]" the sense out of itself by beguiling it (536–37).

The Lady's song also has a similar unselving effect, not on the senses, but on the atmosphere. The Attendant Spirit reports, using the same language of larceny and displacement, that her voice

> stole upon the Air, that even Silence
> Was took e're she was ware, and wish't she might
> Deny her nature, and be never more
> Still to be so displac't.
>
> (556–69)

As we have seen, the silent night serves Comus's purposes not by extinguishing the sounds and sights that might disturb the Lady, but by muffling and concealing them in its feathery thickness. The spiritual nature of sound (its incorporeality for Bacon) allows it to hang in the "spungy ayr" without being dispersed and to intermingle with the spells and enchantments that Comus has lodged there. Air, then, is appropriated by and dedicated to evil especially as it is disposed in the cloudy, dark wood of *Comus*. In reclaiming the air, the Lady's song engages it as Circe's music approaches the sense, catching it unawares. But this process is depicted as an act of liberation rather than one of enslavement, which is the goal of Circean acoustics. The Lady displaces the silent air with a beautiful and wholesome sound rather than leaving it empty. She helps it cast off its enforced silence and become the "vocal air" once again (246).

The effect that divine acoustics has on perceivers is antithetical to that of Circe and Comus's sounds. Rather than depleting the listener of spirit and dulling his or her senses, divine sounds infuse the body with spirit and sound—the two being virtually identical when the sound is harmonious—sharpening the listener's senses and fortifying them with the perspicacity of heaven. The stark contrast between Comus's emotional response to the Lady's singing and the state of ecstasy associated with his mother's music may be seen, for instance, in his comment that he has never heard "such a sacred, and home-felt delight, / Such sober certainty of waking bliss" as the sound of the Lady's voice (261–62). Since the delight the Lady's voice inspires is "home-felt," that is, sensed by the body, Comus is certain that he is awake and sober, not carried out of his senses, floating above the body in an ecstatic trance or dream. In recalling his initial reaction to hearing the Lady's song, the Attendant Spirit affirms that divine sound enlarges rather than depletes the spirit. His sense of hearing grows under the influence of her music to the point where his whole being consists in aurality. "I was all eare," he tells the brothers, "[a]nd took in strains that might create a soul / Under the ribs of Death" (559–61). The extreme augmentation of his aural faculties seems a consequence of the fact that he "took in strains" that replenish his spirit and are capable, as the music of Orpheus was, of creating a soul. Milton appears to follow Ficino, who advises that music is the best way to nourish the spirit, which, he thinks, gets worn out and consumed by mental and bodily activity.[72] Milton also anticipates his own depiction in *Paradise Lost* of angelic spirits, who may choose to live "[a]ll heart . . . , all head, all eye, all ear, / All intellect, all sense" as they please (6.350–51). To become "all sense," then, would seem to be nothing like the fate suffered by Comus's victims, who "roule with pleasure in a sensual stie" (72). Comus and his rout seek pleasurable sensation *outside* of the body, in odors, wine, sweets, starry influence, dance, and games—objects and actions that, when wrongly used, dazzle without invigorating the spirit, much like Comus's spells "cheat the eye" of the truth, giving it less than it deserves (155).

If "carnal sensuality" is the fleshly body's enjoyment of that which deludes the mind and deprives the spirit, then perfect sensibility is the fruit of a robust spirit whose telos is to be uninterrupted by flesh (474). The elder brother gives a description of this heightened state in his encomium on chastity. When someone is found to be truly chaste, as the elder brother explains,

> A thousand liveried Angels lacky her,
> Driving far off each thing of sin and guilt,
> And in cleer dream, and solemn vision
> Tell her of things no gross ear can hear,
> Till oft convers with heav'nly habitants
> Begin to cast a beam on th'outward shape,
> The unpolluted temple of the mind,
> And turns it by degrees to the souls essence,
> Till all be made immortal.
>
> (455–63)

Readers typically respond with skepticism to the elder brother's exorbitant Neoplatonism and optimistic view of the defensive power of chastity. But dismissing this speech as parody, or mere grandiloquence, ignores its cogent representation of aural perception and the economy of spirit in *Comus*. Bearing in mind that the elder brother describes a "[s]aintly" state attainable only if one is judged by heaven to be sincerely chaste, we can learn from this albeit idealized description about the process of evolving into a more refined, spiritous self (453–54).

The refinement of the soul as portrayed in the passage is chiefly initiated by aural sensation. If the ear is spiritual not "gross"—if it is guarded against sounds that will undermine its very functioning—it can be the gateway to higher understanding and immortality. The chaste soul, according to the passage, possesses a spiritual body similar to that of the Attendant Spirit as he hearkens to the Lady's voice and becomes "all eare." Like the extraordinary men in the *Prolusions*, whose surpassing learning gave them souls able to hear the music of the spheres and converse with the gods, *this* soul through "oft convers with heav'nly habitants" gains the aural capaciousness of a spirit. By perceiving exquisitely beautiful sound, a mortal becomes all spirit—a spirit, all ear.

The Lady takes up Milton's implied comparison between the soul and the organ of hearing when she refuses to explain to Comus the doctrine of virginity. She wants to defend the power of chastity but stops and earnestly considers what would come of this. Speaking to Comus (but also thinking aloud to herself) the Lady reasons

> Thou hast nor Eare, nor Soul to apprehend
> The sublime notion, and high mystery
> That must be utter'd to unfold the sage
> And serious doctrine of Virginity,
> And thou art worthy that thou shouldst not know
> More happiness then this thy present lot.
> . . . .
> Thou are not fit to hear thy self convinc't.
> (783–91)

She insists that Comus has "nor Eare, nor Soul," not in order to distinguish these terms, but to emphasize that both are required for and mutually engaged in the act of hearing. Comus's soul has grown "clotted by contagion," "imbod[ied]" and "imbrut[ed]," such that he has lost the spiritual part of himself that allows him to hear (467–68). We are reminded of Bacon's doctrine that "the pneumatical part, which is in all tangible bodies, and hath some affinity with the air, performeth in some degree the parts of the air" as a conductor and medium of sound.[73] The Lady's words will not resonate well with Comus as he is low on these pneumatical spirits, which are integral to acoustical conduction. While the first part of the Lady's argument holds it futile to educate Comus, her next point backtracks a bit, imagining that somehow the sound of her voice may get through to him anyway. She worries that he will literally "hear . . . [himself] convinc't." Just by being in earshot of her divinely sanctioned voice, even without understanding it, a bystander like Comus might find himself convinced by its sound and transformed by an infusion of irresistible spirit. The Lady decides that Comus is not worthy of this gift and that she should like him to stay abject and deaf. Speaking, after all, takes effort and depletes one's precious store of spiritual wherewithal.

In *Comus*, personal agency can do only so much to preserve the spirit from succumbing to environmental evil. Neither physical command of the voice, nor ministering to one's spirit, can sufficiently counter the subterfuge of sinister magic and sin. Throughout the masque the children need certain spiritual supports, beyond their capacity to listen and speak, in order to prevail. The Lady, for instance, calls on "pure-eyed Faith, white-handed Hope" to banish tempting thoughts from her mind, and the elder brother declares that chastity, his sister's "hidden strength," will win her the protection of "[a] thousand liveried angels" (212, 454). At the end of the masque when the characters' personal resources falter, stronger external supports—less connected to the actions of the children—are needed to deliver them from danger. The Attendant Spirit and Sabrina are two of these lifelines. With the protection of a magical herb provided by the Attendant Spirit who is disguised as Thyrsis, the brothers chase Comus and his rout away but fail to liberate their sister from the paralysis that Comus has imposed on her (817–18). A solution is arranged by the Attendant Spirit who calls on Sabrina, the goddess of the Severn, to release the Lady from the clasping charm. Summoning her ritualistically with a song and what he calls "the power of some adjuring verse," his methods exceed the boundaries of Baconian acoustics (857). The utterance of verses signals the mysterious operation of ceremonial or spiritual magic. Yet the invocation is also depicted as a kind of experiment, the likes of which are found in *Sylva Sylvarum*. Bacon holds it possible for sound to carry through the medium of water, and to demonstrate this fact, proposes listening for a sound when the anchor of a ship is let down or when one stone strikes another beneath the water.[74] He also observes that in the myth of Hylas's abduction by the Naiads, when the boy is submerged beneath the water, his voice is still audible to Hercules, though greatly muffled.[75] Similarly, Sabrina and the audience are repeatedly commanded to "[l]isten" while the Attendant Spirit beckons to the goddess where she is "sitting / Under the glassy, cool, translucent wave" (859–60). The successful experiment by which the sound of the Attendant Spirit's voice pierces the water and appeals to the deity indicates the cooperation of the natural laws investigated by Bacon with the spiritual qualities of sound and incantation emphasized by figures of the occult philosophy such as Ficino and Agrippa.

The Lady's passive state at the end of the masque has always generated questions for readers. What is the significance of the spell that renders the Lady unable to move? And why does her silence, which seems to commence when the spell takes hold, continue through the end of the text even after her limbs are brought back to life? The common view is that the Lady's passivity during these final scenes manifests a state of moral impeachment: her paralysis and silence censure her for being overly confident and proud. However, this chapter's analysis of aural perception in the masque suggests that other causes lie behind these famous cruxes. It is necessary to explain the genesis of the Lady's paralysis before interpreting its significance. The masque is clear that the immediate reason for her immobility is that she has been placed her under a spell. Comus warns of this fate when he admonishes the Lady not to move or else he will change her into a statue with his wand (660). At this point she declares that her mind is still free but acknowledges that her body has been "immanacl'd" (664).

In summoning Sabrina to help dissever this "clasping charm" or "the numming spell," Thyrsis confirms that Comus's sorcery has transfixed the Lady in her chair (852).

As to the reason the Lady is rendered speechless and motionless, consider how the masque consistently portrays the categories of assailant and assailed. It represents the Lady's captor as able to overpower her. Thyrsis tells the Elder Brother that Comus's "bare wand can unthred thy joynts / And crumble all thy sinews" (614–15). Thus, the misfortune that befalls the Lady—meeting the duplicitous Comus and falling prey to his magic—cannot reasonably be construed as an indictment of her actions. That Milton took a sympathetic view of the Lady's plight is evident from the fact that he draws a clear parallel between the Lady and Sabrina, the latter of whom he carefully portrays as a spotless paragon of virginity. Thyrsis establishes the parallel, claiming that the river goddess will empathize with the damsel because the Lady is a virgin in "hard besetting need" such as Sabrina "was her self" (855–56). The comparison rings true. Both young women are vanquished by dangerous foes (in Sabrina's case, by her murderous stepmother) and require resuscitation by female spirits (825–41). Thyrsis's association of Sabrina with "innocence" and his insistence that she is "guiltless" and a "[v]irgin pure," despite falling victim to Guendolen's rage, implies that the Lady is similarly unblemished, even though she is stricken by a sinister curse (825, 828, 830). We are not, therefore, meant to interpret the Lady's enchantment as a sign of being singled out for judgment. After all, she demonstrates her moral rectitude in stubbornly refusing to drink from Comus's cup.

With its representation of the wood as a climate sealed off from the world and poisoned by its wicked tenants, the masque suggests that the Lady's predicament is born of a more universal judgment manifested in the very environment. Reduced from full personhood into an inanimate body, the Lady's plight exaggerates the dimming of the children's corporeal senses as they struggle to pierce the wood's darkened, fogged up, and enchanted atmosphere. The masque represents the effect of this depraved climate on the unwitting traveler not in order to uncover and accuse her of her particular sins, but to demonstrate a more general consequence of humanity's collective sin: the difficulty of rightly exercising one's hearing, speech, and other faculties in the face of spiritual evil imbedded in the environment. In her woeful condition, deprived of movement, the Lady embodies the weakness of the sensible body against the worst consequences of the Fall.

Since the Lady remains silent even after the spell has been lifted, readers have often viewed her as a tragic or morally flawed figure whose ability to speak is permanently impaired. However, the masque's depiction of the relationship between bodily spirit and aural phenomena suggests that the Lady's silence is strategic, rather than enforced. She uses temperance—the ethic she praises in her argument for conserving Nature—to govern her decision-making about speaking and listening, saying only what needs to be said and closing her ears to dangerously superfluous chatter. In an effort to preserve her spiritual core, for example, her ears are "unattending" when they meet with flattery (Comus's breath is "ill . . . lost" on her) (272–73). Later, when she considers whether to expound the doctrine of virginity, she pauses to ask (as if to avoid excess in her own speech): "Shall I go on? / Or have I said anough?" (778–79). Her instinct, she says, is to speak out against vice, "[f]ain would I somthing say"; but she reckons the price of declaring what "must be utter'd" in order to defend virginity is too much to pay for the slim possibility of convincing such a hopeless, underserving interlocutor as Comus (782–88). Awaiting uncertain rescue and unable to sate herself with Comus's proffered cup, the Lady's decision not to vocally explicate the divine doctrine protects her store of spirits—we might call it energy—which would be wasted in an attempt to convert Comus's damned soul.

Setting aside the Lady's hesitance to speak too much, there is another knot to untie: why does she sustain her silence after Sabrina lifts the spell and presumably restores her ability to speak? Perhaps readers have attributed more mystery to this crux than it really deserves. Her silence appears to be a mere matter of choice, and it is a wise choice at that. We must ask ourselves, if the Lady deliberately tempers her speech to keep her spirits from waning *before* she becomes like a statue, why should she not follow the same rule, applying even greater caution, *after* she is reanimated and while danger still lurks abroad? I have already noted that the Attendant Spirit strongly discourages the Lady from making any noise ("[n]ot a waste, or needless sound") until they "com to holier ground" (942). Suspecting that while they remain in Comus's haunts they are vulnerable to "som other new device," the Attendant Spirit not only is anxious for the Lady to conserve her flagging spirits, but also worries that her exceptionally beautiful voice—which attracted Comus in the first place—may lure him back (940). This arrangement, wherein the Lady stays silent while the Daemon helps her and the brothers "fly this cursed place," may seem paternalistic to our modern sensibilities—an infringement or erasure of the young woman's identity (938). But the Lady herself seems to have anticipated the danger of speaking out again in this perilous environment and chooses to adopt silence as a protective measure well before the Attendant Spirit suggests this course of action. Righteous in asserting control over her own safety, the Lady's clinging to silence is not a commentary on her moral character, but an illustration of how experience and adaptation are necessary to navigate the hazards of speaking in a corrupted environment.

The modest scope and ceremonial purpose of Milton's masque prevent its serious engagement with broad eco-theological questions outside of the topics of hearing and speech. We must turn to *Paradise Lost* and *Paradise Regained* to discover, for example, what happens when a hellish climate, like that of Comus's wood, bursts open and leaks into a purer world, or how a human being not only resists corruption in the environment, but also shows the way to eradicating it. In representing the Lady's dilemma over how to defend herself, however, the masque shows the seeds of these ideas taking root. Recall that before the Lady falls silent, she indicates that were she to express herself fully, her divinely sanctioned voice and doctrinal authority would destroy Comus's magic and wipe his habitation from the face of the earth (792–98). She opts not to use this ultimate verbal weapon, I argue, because it would expend all her strength, even her humanity. In defeating Comus she would be become what he initially assumed her to be: a protector or genius of the wood. As with the Severn's Sabrina, or Lycidas, the Genius of the Shore, becoming a local guardian entails dying and being made immortal—something that is hinted at by the Lady's temporary immobilization. In rehearsing this possible, but ultimately unfulfilled, resolution to the masque—reclamation of the enchanted wood through martyrdom—her paralysis points to the kind of heroic action Milton deems necessary for eliminating corruption from the climate. Defending and speaking the truth to the bitter end, is an Orphic deed that permeates and galvanizes the environment, moving "dumb things" and "the brute Earth" to shake off their parasitical possessors (795–96). Happily, for the Bridgewater children, this masque has a pre-set ending: the Lady must be safely delivered to her parents, and therefore Milton takes over the task of reclaiming the dark air of the enchanted wood by transforming the scene into a festive dance.

Since the magic of genii and daemons—the benign counterpart of Comus's sorcery—is necessary to revive the Lady and dispel the incapacitating fogs and enchantments of the wood, Milton's masque cautions that even when the senses are properly guarded, they can be overpowered by "felonious" influences in the environment. Experimentation and temperance with regard to hearing and speaking are powerful tools of moral preservation. But an epistemology such as Bacon's that ignores the vulnerability of the acutely aural soul to potential corruption seeded in the atmosphere is in Milton's estimation deeply flawed and calls for supplementation. As we turn in the next several chapters to the epic poetry, we shall see that influences from classical meteorology, occult philosophy, and the Christian tradition remain crucial complements to Milton's interest in the era's increasingly sophisticated, often mechanistic accounts of atmospheric phenomena. Whereas mechanistic models of air phenomena offer opportunities to depict corruption, the older assumption that human and world spirit are fundamentally alike is critical to showing that the boundary between subject and climate may be easily and devastatingly transgressed. Exploring the subject of the Fall in the epic *Paradise Lost* enables Milton to explicitly identify the original source of atmospheric corruption. Turning from Milton's early works to his late masterpieces, we see a shift in the poet's focus with respect to the atmosphere from exploring its status as a damaged medium of aural expression to portraying the fallen atmosphere as indelibly changed by the Fall, corrupted by the presence of evil spirits, and manifesting a perpetual sign of humanity's original sin.

CHAPTER 3

The Power of the Air in Milton's Epic Poetry

Of all the hazards of fallen existence represented in *Paradise Lost,* the air poses the most pressing threat to life. Its potential antipathy prompts the Son's first compassionate act after judging Adam and Eve: "[T]hen pitying how they stood / Before him naked to the air, that now / Must suffer change . . . he clad / Their nakedness" (10.211–13, 216–17). Adam soon realizes that this divinely provided and mysteriously symbolic clothing, though it shows the way, is not sufficient to protect them from the elements (10.219–23). So pitiless will the air become that they must plead for additional relief:

>   if we pray him, will his ear
> Be open, and his heart to pity incline,
> And teach us further by what means to shun
> The inclement seasons, rain, ice, hail and snow,
> Which now the sky with various face begins
> To show us in this mountain, while the winds
> Blow moist and keen.
>
> (10.1060–66)

The word "inclement" echoes an earlier description of the tempestuous sky above "[t]he Paradise of Fools," where vain, heretical, and superstitious souls are exiled after life in this world (3.426, 496). The skies over these earthly and superlunary "Paradise[s]" are inclement in the usual sense, producing the opposite of what we would call fine weather. But the term has moral implications as well. Where clemency implies the merciful or gentle use of power—"it droppeth as the gentle rain from heaven"—inclement weather is relentless in producing human suffering.[1] The cruelty of the weather manifests in the description of Limbo's atmosphere where hooded clerics, hopeful of gaining heaven, become the "sport of winds," blown from its door "[i]nto the *devious air*" (3.493, 489; emphasis added). With judgment mediated by the very air surrounding them, Adam and Eve's only hope for protection is divine clemency.

Another way to say the air is "devious" is to say that it errs. Milton is aware of the homophonic relationship between "air" and "error" and the union of these concepts in Satan's well-known title as "the prince of the power of the air" from Ephesians.[2] Milton's epic illustrates how the air becomes wayward, or materially and spiritually biased toward evil and cruelty, by recounting how Satan rises to power as its prince. In the process of assuming his lordship over the air, Satan becomes implicit in it; his deviousness inhabits it. In portraying the air as charged with demonic essence, Milton expresses his vitalist animist notion of the universe as well as contemporary beliefs about the spiritual origins of the weather. The ominous figure employed by Adam to personify the air's mutability, its "various face," recalls the series of troubled expressions that contort Satan's countenance as he resolves to deceive man (4.114–17). Satan's facial contortions are writ large in postlapsarian weather, and, conversely, postlapsarian weather is prefigured in his face, expressing the bodily participation of demonic power in the air.

Karen L. Edwards, Joanna Picciotto, and John Rogers have detailed significant connections between Milton's poetry and seventeenth-century science, but Milton's fascination with the weather has been overlooked—and understandably so.[3] Meteorology is routinely excluded from "the historiographical tradition that describes the changes that natural philosophy underwent during the early modern period."[4] Yet meteorological forces permeate the material universe of *Paradise Lost* and are instrumental to its moral action. Scholarship has underestimated the pertinence of meteorology to the epic possibly because modern notions of the weather exclude phenomena that in Milton's era were deemed meteorological. The fallen angels' movements in hell trigger mineralogical processes that, from our perspective, appear wholly unrelated to meteorological phenomena such as mist, dew, and rain and are even further removed from astronomical events such as comets and shooting stars. Yet the boundaries between these apparently distinct explanatory paradigms were at best blurry in the late Renaissance, as this sixteenth-century definition of meteorology demonstrates: "It is the part of physics that is concerned with what comes to be in the regions of the air or in the belly of the earth."[5] To us, the atmosphere is the primary source of the weather. But the term "atmosphere" was not coined until the mid-seventeenth century, and the region to which it refers forms just a part of the weather system as envisioned by early meteorology.⁶ Rather, the far more diffuse concept of "air," a substance composed of hot and cold fumes thought to penetrate beneath the ground, subtended all meteorological activity in the world.

The modern notion of air as a neutral mixture of gases lies far off from what Milton would have understood about the air he breathed. According to the *OED*, the primary definition of the noun "air" in Milton's time was an "atmosphere contaminated by noxious fumes, vapours, etc." or "such contaminating fumes themselves."⁷ Pestilence was not thought of as a pollutant, to use a modern term, but a defining feature of the air. Recognizing that the air was understood to be malignant by default allows us to trace seemingly unrelated events in *Paradise Lost* to a common demonic source. By endowing the demons with meteorological bodies as they gradually lose their angelic ones, Milton provides an account, consistent with scripture and contemporary attitudes about the air, of why the weather is the primal antagonist of man. This account of the weather, one of several origin stories told by the epic, reveals the immanence of demonic agency in Milton's representation of the natural world and the logic behind his account of Satan's "covert" temptation of man (2.41). Satan's scripturally defined role as "prince of the power of the air" seems to license him to corrupt every stratum of the air, from the climate down to the breath of individual organisms. Access to the voice through the power of the air enables him to interfere secretly with Eve's mind while she sleeps and later to tempt her in the guise of the serpent. Thus Milton mythologizes the ancient association between devils and the air to portray the original devastation of the earth's climate, explain the vicious character of fallen weather, and demonstrate the lethal potential of sound.

Milton's dynamic representation of air is not circumscribed by a single theory or explanatory discipline but rather reflects insights from both occult and empirically oriented traditions. To appreciate his explication of the spiritual sources of weather, it is necessary to consult alternative and supplementary traditions regarding the air, including scriptural exegesis, demonology, and hagiography. Recovering the history of climate degradation woven into Milton's representation of the Fall elucidates the critically neglected association in *Paradise Lost* of the demons with meteorological processes as well as with what I call the "pneumatics of temptation." Satan's special relationship to air empowers the acoustical method he uses to tempt Eve discussed in detail in chapter 4.

Let us begin with Milton's representation of Satan as ruler of the air. Throughout his poetry, Milton affirms the doctrine that in the fallen world, Satan and his demonic crew are allowed to possess the air—particularly its middle region. This is the middle layer (*media regio*) of the sublunary atmosphere, which Milton and his contemporaries divide into three parts: a hot upper region (*suprema*), a habitable lower region (*infirma*), and a meteorologically eventful middle. The relatively cold middle part reaches as high as the tallest mountains and receives vapors from below that gather there, form clouds, and turn to rain.[8] The idea that the fallen angels might settle this region of air emerges during the demonic debates in book 2 of *Paradise Lost* and gradually comes to fruition as Satan infiltrates Earth's atmosphere.[9]

At the beginning of the poem, Beëlzebub beguiles the infernal council with conjectures about the new world, where, he speculates, they may "in some mild zone / Dwell," and whose

> soft delicious air,
> To heal the scar of these corrosive fires
> Shall breathe her balm.
> (2.397–98, 400–402)

Experiencing the earth's atmosphere for himself like a sailor "[w]ell pleased" with its exotic fragrance reinforces Satan's plan to resettle hell's inmates there (4.164). He tells the angelic guard that he aims either "here on earth, or in mid-air" to install his "afflicted powers" (4.939–40). After the Fall, Satan proclaims that his offspring, Sin and Death, should "on the earth / Dominion exercise and in the air" (10.399–400). Then, in *Paradise Regained,* we learn that at least some of hell's other citizens have since migrated to Earth and now inhabit "the middle region of thick air" (2.117).

One might dismiss Milton's association of the devils with the air as merely an allegory or a parody of religious superstition. Their promised habitation shares the attributes of impermanence and airiness with Milton's Ariostian Limbo, where he locates the souls of "[e]mbryos and idiots, eremites and friars" and "all things transitory and vain," which "like aërial vapors flew" (3.445–46, 474). Descriptions of the realm of air occasionally display the implausible concreteness and incongruous detail that Mindele Anne Treip associates with Miltonic allegory.[10] Satan's "place" in *Paradise Regained,* for instance, is characterized anachronistically as a "gloomy consistory" among "thick clouds and dark tenfold involved" (1.39, 41–42). Yet Milton's airy devils are not mere epic machinery, as are the sylphs of Alexander Pope's *The Rape of the Lock* (1712). Milton may mock the presumed loftiness that the demons derive from their airy station, but nowhere does he question the scriptural basis of their association with air. That demons occupy the air under Satan's headship is presented as a fundamental condition of the postlapsarian world. We see this in book 10 of *Paradise Lost,* when the narrator describes the life and Resurrection of Christ as the fulfillment of the serpent's curse:

> So spake this oracle, then verified
> When Jesus son of Mary second Eve
> Saw Satan fall like lightning down from heaven,
> *Prince of the air;* then rising from his grave
> Spoiled principalities and powers, triúmphed
> In open show, and with ascension bright
> Captivity led captive through *the air,*
> *The realm itself of Satan long usurped,*
> Whom he shall tread at last under our feet;
> Even he who now foretold his fatal bruise.
> (10.182–91; emphasis added)

The words "long usurped" imply the stability and historical truth of the doctrine of Satan's power over the air, as does Satan's characterization of the air in *Paradise Regained* as "[t]his our *old conquest*" (1.46; emphasis added). The passage alludes to multiple scriptural verses relating to Satan's airy kingdom: "the prince of the power of the aire, the spirit that now worketh in the children of disobedience"; "I beheld Satan as lightning fall from heauen"; "having spoyled principalities and powers, he made a shew of them openly, triumphing over them in it"; and "[w]hen he ascended up on high, he led captivitie captive."[11] Milton's allusion to these verses at this grave moment in the poem—God's judgment of Satan in the serpent—demonstrates that Milton, whose theology is scrupulously and minutely calibrated to scripture, insists on the authenticity of the doctrine of Satan as "prince of the power of the air." Allusions to many of the same biblical passages in Michael's description of the Ascension ("[Christ] shall ascend / With victory, triúmphing through the air" and "there shall surprise / The serpent, prince of the air") confirm that Milton considered this doctrine a crucial part of revelation, basic to understanding Christ's mission on earth and the nature of the evil he overcomes (12.451–54). In *De Doctrina,* Milton cites Ephesians 2:2 in conjunction with other passages to show that bad angels wander "the whole earth, the air, and even heaven" and that God suffers these wanderings in order to carry out his judgments (*CP* 8.1:333, 353, and 433).[12] By placing demonic activity in the world under divine government, Milton affirms the reality of Ephesians 2:2. God not only permits Satan's occupation of the air, but he also expresses his will through it as an instrument of divine providence.

CHURCH DOCTRINE AND METEOROLOGY

The sometimes shockingly heretical Milton is hardly heterodox in the literalism of this belief, however odd it may seem to us now. He echoes a truism of the premodern world: demons inhabit the air and exercise control over meteorological phenomena such as the weather. The writings of the Church Fathers shed light on the origins of this belief. In a gloss on Ephesians 6:12, for instance, St. Jerome invokes "the view of all teachers that the air which divides between heaven and earth and is called empty space is full of contrary powers."[13] Jerome's sense of the ubiquity of the doctrine likely refers to its acceptance outside of, as well as within, Christianity. Clinton E. Arnold argues that the intent of the epistle to the Ephesians and its verses pertaining to the "prince of the power of the air" is to offer relief in the form of Christ's salvific power to a community whose magical religious practices reflected their belief in the real agency of spiritual beings.[14] Ephesians consolidates these pagan "powers" under one enslaving prince and explains that he along with his host of demons have been subordinated to the power of Christ.[15] The figure Christ overcomes is possibly depicted as a ruler of the domain of air because "the air was regarded as the dwelling place of evil spirits in antiquity."[16]

In *The City of God against the Pagans*, Augustine alludes to a non-Christian tradition that conceived of the air as full of supernatural beings, and he relates it to Christianity. In book 10, chapter 21, headed "The measure of power granted the demons for the glorification of the saints through their strength to endure suffering; for the saints triumphed over the spirits of the air, not by doing their pleasure, but by abiding in God," Augustine compares the martyrs of the church with the Greek heroes, the fabled "souls of the departed as earned distinction to some degree in this life."[17] Augustine says their name derives from Hera—the ruler of the air according to Greek myth—and they dwell in the atmosphere with the demons.[18] The martyrs of the "true" religion, however, deserve the name of hero for another reason: "not because they and the demons could be members of one community in the air, but because they overcame these same demons, that is to say, powers of the air, and in company with them Juno herself. . . . Our heroes, if usage permitted us so to call them, do not in the least resort to gifts to overbear Hera, but to valour that comes from God."[19] Augustine does not contest the existence of deities and demons of the air, but he stresses that the Christian orientation toward them is hostile, not conciliatory. Milton's portrayal of the middle air in *Paradise Lost* aligns with this account, where the air functions as an ecumenical space for demons. Just as Augustine acknowledges that demons of the air are real to both pagans and

Christians, Milton's middle air offers asylum to all adversaries of the Judeo-Christian God regardless of their genealogy (1.515–17).[20]

Medieval and Scholastic writers applied the doctrine of a demonic realm of air, lorded over by Satan, to the field of meteorology, alleging that these diabolical powers could influence the weather and use it to torment or punish men.[21] They were not first to attribute meteorological phenomena to supernatural or divine agency. While the Greek natural philosophers, beginning with Thales of Miletus (ca. 620–ca. 546 BCE), eschewed supernatural explanations in their rational accounts of meteors, the ancient poets famously assigned the motions of the atmosphere to individual gods, such as Zeus, god of thunder and lightning, and Aeolus, lord of the winds. In the Christian era, medieval and Renaissance authorities frequently relied on scripture to explain atmospheric marvels; they often viewed rare occurrences such as comets as portending future events and attributed weather patterns to the operation of Providence.[22] It was especially common for professors at the Lutheran universities to interpret such meteorological activity as a sign of God's will. Martin Luther himself was convinced of their predictive or providential significance, and Philip Melanchthon, who created much of the schools' curriculum on natural philosophy, insisted on the moral and historical meaning of disastrous weather.[23] Thus, throughout antiquity and into the Renaissance, there existed a thoroughly teleological tradition of meteorological explanation, and within this tradition, churchmen, occultists, and natural theologians speculated that catastrophic or rare weather events—storms, thunderbolts, lightning—were the work of evil spirits.[24] Thomas Aquinas's avowal that, "with God's permission, demons can induce turbulence of the air, stir up winds, and make fire fall from heaven," reflects the solemn orthodoxy of the idea; but it was also taken up by unorthodox authors such as Jean Bodin, whose *Universae Naturae Theatrum* (1596) was deemed heretical enough by the Catholic Church to warrant censorship throughout the Counter Reformation.[25] A curious amalgam of religion and science with a metaphysics similar to Milton's (replete with corporeal angels, demons, and souls), the *Theatrum* attributes violent winds and storms to demons "acting under divine command."[26]

The notion that demons rule the air appears in hagiographical traditions as well. The legend of St. Anthony, as told by Athanasius in the fourth century, inspired a number of Renaissance artworks that depict the sky as the territory of devils. Several focus on Anthony's vision of himself being carried into the air and taken to task by "foul and terrible figures standing in the air."[27] Athanasius writes, "[Antony] was amazed to see how many foes our wrestling involves, and how many labors someone has in passing through the air, and he recalled that this is what the Apostle said, *following the prince of the power of the air.*"[28] In his first known painting, *The Torment of Saint Anthony* (ca. 1487–88), a young Michelangelo depicts the saint as being attacked in midair by winged, scaled, and club-wielding demons.[29] A triptych by Netherlandish painter Hieronymus Bosch (ca. 1450–1516) of Anthony's temptation and an etching of the same subject by Jacques Callot (1592–1635), a printmaker from the Duchy of Lorraine, present skies that swarm with evil spirits in various shapes and horrifying postures.[30] In one panel of Bosch's *Temptation* (ca. 1500), Anthony is lifted into the air and tormented by demons in flying ships. The painting *Sea Storm* (ca. 1508–28), by Palma Vecchio (1480–1528), also features demonic sailors.[31] In Vecchio's piece, which represents the fourteenth-century legend wherein St. Mark, St. George, and St. Nicholas rescue Venice from a great storm, the saints steer a small craft toward an ominous black ship that has been commandeered by devils.[32] Shadowy silhouettes on the riggings of this bark blend into the stormy atmosphere around the boat, giving the impression that the demons are embodiments of bad weather.

Early modern viewers would have easily decoded these motifs, all of which present the devils in their traditional and scripturally warranted role, harnessing the power of air to execute punishments on man. Their frequent association with airborne ships possibly inspired the many nautical analogies applied to Satan in *Paradise Lost,* such as the comparison of him to a fleet of mercantile vessels that "[h]angs in the clouds, by equinoctial winds" (2.637). Nautical imagery lends itself to demonological representation perhaps because the figure of a ship evokes mobility, agility, and worldliness—traits that enable aerial spirits to harass and misguide mortals. Milton's suspension of the boats in midair—the fleet "[h]angs in the clouds" much like the airborne vessels of Bosch's *Temptation*—signals Satan's supernaturalism and underscores his affiliation with the atmosphere. Completely dependent on wind for movement, a sailboat is both an instrument of weather and its captive, expressing and channeling the wind's motion while continually in danger of being overpowered by it. As a metaphorical vehicle for Satan, the sailboat thus reflects his paradoxical status as a power of air: he is permitted to direct and exploit the power of the atmosphere, but is ultimately reliant on God for propulsion and, indeed, for ontological continuance. As we shall see in chapter 5, the discovery of air pressure and the development of modern weather instrumentation also contributed to Miltonic analogies between satanic flight and the experience of being tossed on a churning ocean.

Meteorological agency is not reserved exclusively for demonic beings. Luther is purported to have said, "Winds are nothing but spirits, either good or evil. The devil sits there and snorts, and so do the angels when the winds are salubrious."[33] Aquinas likewise holds that both good and bad angels (*spiritus boni sed etiam mali*) can impel local atmospheric changes, but only insofar as God will allow.[34] Milton's God is similarly indifferent to the moral status of his meteorological agents. According to Raphael, either band of angels, good or apostate, can "wield / These elements, and arm him with the force / Of all their regions" (6.221–23). The climate changes accomplished by good and evil meteorological agents after the Fall, however, reveal a key point of distinction between them. Sin and Death, for instance, cause damage to the stars and planets simply by going near them:

> Their course through thickest constellations held
> Spreading their bane; the blasted stars looked wan,
> And planets, planet-strook, real eclipse
> Then suffered.
>
> (10.411–14)

The simultaneously active and passive word "[s]preading" suggests that they impart their fatal nature automatically, if not compulsively, to whomever and whatever crosses their path. The good angels interfere with the same apparatuses of weather in an entirely different way. They are depicted as "prescrib[ing]" the planets' movements and teaching the fixed stars when to cross man (10.657). Unlike the ambiguous term "[s]preading" applied to Sin and Death, these verbs present the angels' approach to the weather as deliberate, detached, and controlled. This difference in diction reflects the agents' volitional position with respect to God. The angels willingly fulfill His commands, whereas the "dogs of hell" seem to do so instinctively, unwittingly executing God's plan by subjecting the world to their pernicious influence (10.616).

The ambiguous language used to characterize Sin and Death's contact with the heavens possibly reflects their status as allegorical beings; it also fits with a general pattern by which the poem elaborates their association with air. Unlike the good angels, who keep their individuality distinct from Nature by transforming it didactically via "precept," the demons physically assimilate themselves to the climate and vice versa (10.652). Their immersive agency is evident from the fact that "bane" flows from them like a fluid substance into the heavens; it is also evident from the reflexive syntax of "planets, planet-strook," which captures how the monsters reflect and embody the planets' malignant function. Throughout the poem, Milton not only employs figures of mixture to insinuate the devils' environmental incorporation, but he also implies their likeness to meteorological forces analogically. We see both techniques in the alternative views of Satan bubbling up through a fountain in Paradise, literally "involved in rising mist," versus his stalking the serpent figuratively, "[l]ike a black mist low creeping" (9.75 and 9.180; emphasis added). The duality of these representational modes creates a flickering effect, projecting an image of the devils as both embodiments and reflections of the weather.

THE CLOUDING OF THE SATANIC BODY

Raphael's narration of the battle in heaven refers often to Nature's elements (wind, thunder, clouds, rain, hail, dew, and mist) to describe the angels' feats of war. As it applies to the good angels, Raphael's naturalistic imagery simply reflects and underscores the sublimity of their deeds. The tempest said to accompany the stroke of Abdiel's sword and the simile that compares the impact of the blow to a subterranean wind dislodging a mountain, for instance, impute cataclysmic force to Abdiel (6.190–97). Raphael invokes the power of wind to characterize Michael's fierce combat with Satan, and the supreme powers of thunder and lightning are reserved for the omnipotent Father and Son (5.893, 6.308–10, 836, and 849). When Raphael applies meteorological figures to the apostate angels, however, there is a corresponding change in tenor. Rather than aggrandizing Satan's faction, the imagery presents the apostate angels as physically darkened or burdened by pain, foreshadowing their deteriorated condition as vanquished spirits of hell. Cloud imagery plays a key role in portending this fall.

Neither dangerous nor remarkable in themselves, clouds are harbingers of inclement weather. Insofar as they characterize intellectual beings in Milton's works, they also perform a semiotic function by manifesting a person's psyche in the cloudiness of his or her face. The characterization of divine persons with cloud imagery poses a special case. The presence or absence of clouds surrounding God is not so much a reflection of his inward state as the divinity of those who view him. Thus, God appears to the angels, who are too weak to look at him directly, "through a cloud," but addresses the Son "without cloud, serene" (3.378 and 11.45). And like a magical glass, the Son's serene face reveals God to man in a form he is capable of seeing; in his "conspicuous countenance, *without cloud* / Made visible, the almighty Father shines" (3.385–86; emphasis added). With creatures, however, clouds do not have a shielding or veiling function, but signify, rather, the presence of sin, injury, or inner turmoil. When he wakes Eve from her bad dream, for instance, Adam banishes clouds of distress from her visage: "Be not disheartened then, nor *cloud* those looks / That wont to be more cheerful and serene" (5.122–23; emphasis added). Because he believes—

perhaps naively—that Eve has not been corrupted by the dream, Adam thinks her face should be free of clouds. In *The Second Defense of the English People* (1654), Milton uses the phrase "without a cloud" to describe the deceptively uninjured appearance of his blind eyes.[35] Here, the absence of clouds, signifying healthy organs of sight, gives a false reading of the inner condition. As with real weather prediction, Milton seems to point out, looks can be deceiving.

The clouds that gather around the embattled rebels in heaven, however, do not misrepresent their mental and physical disquietude. When one of the rebels' leaders, Nisroch, responds to Satan's call for martial innovation, Raphael tells us that he is badly wounded and "cloudy in aspéct" (6.450). Likewise, on the morning of the second day of battle, Zophiel informs the obedient angels of the hazy shape of the approaching foe:

> so thick a cloud
> He comes, and settled in his face I see
> Sad resolution and secure.
> (6.539–41)

Due to the delaying effect of enjambment in these lines, we momentarily imagine that Zophiel has seen a thick cloud settled in the enemy's face and recall the moment in book 4 when Uriel informs Gabriel of the telltale passions he discerned in Satan's countenance (4.570–71). In both cases, facial disfiguration suggests the inner state of misery that accompanies sin.

The fallen angels' suffering is physiological as well as mental. The shocking revelation of book 6 that the rebel angels can feel pain manifests the bodily consequences of sin. Satan insists that his self-healing wound is proof of their "[i]mperishable" form and "native vigour" (6.435–36), and Raphael confirms Satan's immortality, comparing the resilience of angelic substance to the fluid texture of air (6.348–49). But neither account of their injuries addresses the material cause of this susceptibility. It takes the dissolution of their forms—or nearly that when the mountains of heaven come crashing down on them—to discover the corporeal basis of their pain:

> Their armour helped their harm, crushed in and bruised
> Into their substance pent, which wrought them pain
> Implacable, and many a dolorous groan,
> Long struggling underneath, ere they could wind
> Out of such prison, though spirits of purest light,
> Purest at first, now gross by sinning grown.
> (6.656–61)

Their inability to wiggle out of the crushed suits of armor quickly suggests that the viscosity of their spiritual beings has changed. But what has it changed into? The meteorological pun on the verb "wind" invites us to conceive of their now-gross substance in terms of the weather. The rebel angels' "dolorous groan[s]" recall the wailing of a gale, and the coagulation of their pure essence reminds us of Zophiel's association of the marching rebels with a "thick" cloud.

One reason Milton might have used figures of weather, such as clouds and wind, to portray the literal and metaphorical hardening of the disobedient angels is that, from the perspective of Scholastic physics, meteorological phenomena represent an imperfect state of matter. The natural law that Milton's God gives to the elements in their ideal state—to "know / No gross, no unharmonious mixture foul"—implies that their nonideal condition in the fallen world is mixture and disharmony (11.50–51). In this view, he echoes medieval and Renaissance commentators who typically allude to Aristotelian meteorology as the study of "imperfect mixtures."[36] Not all mixtures were thought to be defective, but "[m]eteorological phenomena were considered imperfect because they were composites of the elements that had not been transformed into a new substance."[37]

Matter theory clarifies why the demons' embodiment of the weather accompanies the loss of their refined angelic bodies. The very term for meteorological formation, "imperfect mixture," connotes physical depravity as it directly negates the initial good condition of all God's creatures, who are given "perfect forms," are "[p]erfect within," and who "God made ... perfect, not immutable" (7.455, 8.642, and 5.524). Meteorological mixture, moreover, represents a state of unrest, which is precisely the condition Satan embodies. The elements' endless transformations into one another, outlined in Aristotle's theory of generation and corruption, reflect the tortuous revolution of passions throughout Satan's body.[38] The inner turmoil he experiences before descending into Eden, "[n]ow rolling" and "boil[ing] in his tumultuous breast," reveals the essentially meteorological dynamic of his being:

> each passion dimmed his face
> Thrice changed with pale, ire, envy and despair,
> Which marred his borrowed visage, and betrayed
> Him counterfeit, if any eye beheld.
> For heavenly minds from such distempers foul
> Are ever clear. Whereof he soon aware,
> Each perturbation smoothed with outward calm.
> (4.16, 114–20)

Satan's face is presented as a sky, his mind a turbulent microclimate, and his passions (significantly the narrator enumerates four of them) the intermixing elements. Because his body is subject to such commotions, Satan's face, whether it openly expresses distress or is artificially smoothed over, is a medium of meteorological signification.

It is easy to see why cloudiness, a partial mixture of the elements of air and water, serves as a poetical figure for the fallen spirits' states of impurity. Yet Milton had reason to believe that the demons had actually transformed into the stuff of clouds. By assigning them the qualities of clouds or mist, Milton conforms to the way early theologians discuss the bodies of fallen angels. Augustine locates the rebellious angels in the earthly atmosphere and grants them bodies of air.[39] If they had bodies "of a celestial nature" before their fall, he reasons, then when they sinned they "were changed into the element of air so that they might undergo suffering from the element of fire."[40] The implied logic is that punitive fires may hurt bodies composed of an element such as air, though not bodies of a purer celestial substance, such as those of the faithful angels. Associated with the alteration of the disobedient angels' substance is their ejection from the highest region of atmosphere, which contains pure air "joined in common bond of peace with the heavens," and relegation to the misty atmosphere below, which is "full of water in a refined and vaporous state."[41] In interpreting Augustine, Peter Lombard, author of the influential twelfth-century work *Sentences,* reinforces the implied connection between the angels' elemental change and their atmospheric relocation, concluding that their substance thickened according to the nature of the air in their new habitation: Augustine "seems to say that all angels before their confirmation or fall had aerial bodies, formed from the purer and higher part of the air and suitable for acting, but not for suffering. And such bodies were preserved for the good angels who remained steadfast.... But the bodies of the evil angels in their fall were changed into an inferior quality of thicker air. For just as they were cast down from a worthier place to a lower one, that is, into this cloudy atmosphere, so their refined bodies were transformed into inferior and thicker ones."[42] Here we find a theological precedent for connecting the fallen angels' substance with the misty realm of atmosphere they eventually populate, or, the middle air.

Milton does not identify the fallen angels with a single element, but Augustine and Lombard's precedent suggests why he portrays the corrupt angels as resembling and behaving like clouds and condensing or acquiring thickness. These depictions reify the angels' moral and physical degradation and anticipate their future confinement to a subempyreal region where "there is air, but it is air saturated with the vapor that produces winds when stirred, lightning and thunder when violently agitated, clouds when gathered in a mass, rain when condensed, snow when clouds are chilled."[43]

THE METEOROLOGY OF SATANIC SPEECH

The purpose of Milton's association of the devils with meteorological phenomena is twofold. As I have argued, it illustrates the origination of their role as weather makers in the fallen world and the significance of atmospheric inclemency as a penalty of original sin. Additionally, it supplies an explanation for the nature of Satan's overtures to Eve, which take the form of delusive sounds culled from the atmosphere. The dream Satan conjures pneumatically through Eve's ear, as with an organ or "inspiring" breath and the words he addresses to her through the serpent, depend equally on his flexible aerial nature and ability to simulate sounds (4.804). Milton endows the demons with a physical affinity to atmosphere in part to render their scripturally designated role as powers of air and weather but also to enable Satan's atmospheric interference with voices in Paradise. *Paradise Regained* portrays Satan's ability to channel his essence into aural form as the principal way he misleads men. His "chosen task," according to Jesus, is "[t]o be a liar in four hundred mouths"; he is "composed of lies," which are his "sustenance" and "food" (1.407, 427–29). Recalling Satan's vocalization of the serpent, these characterizations affirm the continuance of Satan's vocal power after the Fall and attribute it to his bodily composition (he is "composed of lies"). We find the archetype of this method of deception, by which Satan channels himself into the mouths of others, in *Paradise Lost*, where Satan transfers his airy substance into the serpent and uses it to deceive Eve.

But why does Satan's meteorological power translate into the ability to counterfeit the speech of others? A partial answer may be found in the language of meteorological theory. As we saw in chapter 1, early meteorological treatises in their frequent invocation of respiratory models imply a fundamental relationship between weather and vocality. This convention derives from Aristotle's theory of exhalations, which attributes diverse phenomena such as comets, metals, and earthquakes to the action of certain vapors above and beneath the earth's surface.[44] Since virtually all seventeenth-century writers, including René Descartes, incorporated exhalations into their meteorological theories, the associated oral terminology is a hallmark of such texts.[45]

For Aristotle, the air itself is comprised of exhalation. He defines it as a dynamic compound "made up of these two components, vapor which is moist and cold . . . and smoke which it hot and dry."[46] Various physical factors, such as the movement of the stars, activate these gases and trigger meteorological events: dry exhalation ignites to produce shooting stars or lightning; moist exhalation condenses into cloud, mist, dew, or rain.[47] Illustrating Milton's vitalistic conception of matter, exhalation in *Paradise Lost* tends to confer orality on all its meteorological processes. Adam and Eve describe the earth's rising "mists and exhalations" as voicing God's praise as they infuse the atmosphere with clouds and rain, and, at the hour of their nuptials, the earth vocalizes her felicity with "fresh gales and gentle airs," an allusion to the popular musical form known as the air (5.185 and 8.515). But the same perfumed breezes, which Milton fits with "odoriferous wings" as if to liken them to winged spirits, "entertain[]" the fiend with their pleasant scents on his arrival to Paradise and "whisper whence they stole / Those balmy spoils" (4.157–59, 166). Their whispers are agreeable and even serviceable to Satan because exhalations, as ingredients of air, are his inheritance. Therefore, in addition to cycling elemental praise through a breathing creation, exhalation exposes meteorological phenomena to demonic appropriation. Wherever there is exhalation, there is opportunity for satanic corruption.

Since human utterance is made from bodily exhalation, Milton portrays dew and condensation, products of meteorological exhalation, as mechanisms of environmental utterance. Because they possess clouded or vaporous bodies, the fallen angels are implicit in natural processes of liquefaction, which transmit their spiritual being into a more corporeal form. The simile in book 5 that compares the defected angels to droplets of morning dew thus represents the rebel angels' swift relocation to the north as a result of condensation:

> Satan with his powers
> Far was advanced on wingèd speed, an host
> Innumerable as the stars of night,
> Or stars of morning, dewdrops, which the sun
> Impearls on every leaf and every flower.
>
> (5.743–47)

Here, the shifts from night to day, aerial to solid, high to low, foreshadow the "foul descent" Satan undergoes when he slinks back into the garden "wrapped in mist / Of midnight vapor" (9.158–59 and 9.163). Just as aerial angels "[i]mpearl" themselves into dewdrops, Satan condenses by "incarnat[ing] and imbrut[ing]" himself into "bestial slime" (9.165–66).

Satan undertakes this moral and material descent from mist to slime to inhabit the snake and make it speak or seem to speak. The idea that demonic words materialize from the air like dew recurs elsewhere. When the narrator describes how Belial, the most lewd and "gross" of all the fallen spirits, speaks, he alludes to the story from Exodus, wherein God feeds the Israelites bread from the sky:

> his tongue
> Dropped manna, and could make the worse appear
> The better reason, to perplex and dash
> Maturest counsels: for his thoughts were low;
> To vice industrious, but to nobler deeds
> Timorous and slothful: yet he pleased the ear.
> (1.491 and 2.112–17)

The conceit caricatures Moses's identification of manna with God's word: "And he humbled thee, and suffered thee to hunger, and fed thee with Manna."[48] It also connects Belial's manner of speaking with the mechanism by which manna was deposited in the wilderness—condensation or precipitation of dew. In Milton's poem, this meteorological process poses an opportunity for latent powers of air to express or convey themselves into a more solid manifestation. Condensation simultaneously signifies devilish articulation and the natural mechanism by which demonic cloud or vapor becomes incorporate.

Although Milton's fallen angels eventually congregate in the midair, they are first expelled into the depths of hell. In addition to assimilating the humid properties of their earthly locus, the devils therefore occasionally exude and embody their infernal environment, as Satan's famous exclamation—"Which way I fly is hell; myself am hell"—poignantly illustrates (4.75).[49] They are capable of reflecting both climates because of their likeness to air, which is composed of hot and cold exhalation. That the devils' meteorological identities persist in hell is evident from the many meteorological figures applied to their infernal activities. Recall the simile comparing the devils' military exercises to portentous wars "[w]aged in the troubled sky" by "airy knights," or the characterization of Sin and Death as "two polar winds blowing adverse / Upon the Cronian Sea," or, more vivid still, the notion that some spirits spend their rage on hell's soil, tearing up its "rocks and hills, and ride the air / In whirlwind" (2.534, 536, 540–41, and 10.289–90).

Milton analogizes hell's mineral-rich core and singed terrain to earth's Tartarean regions and smoking promontories, encouraging us to imagine the devils' torture chamber in geological terms (1.230–37, 684–88, 2.858, and 6.54). We are thus correct to think of hell's ventilating winds as similar to the exhalations believed to suffuse the earth's crust.[50] The meteorological implications of the devils' mining activities and use of metals (e.g., 1.545–49, 670–99), which derive from mineral-forming exhalation and the presence of sulfur beneath the ground, is considered in more detail in chapters 4 and 6. Milton directly invokes the concept of exhalation to portray the material and acoustical qualities of the structure of Pandaemonium, which the devils build from their infernal environment:

> Anon out of the earth a fabric huge
> Rose like an exhalation, with the sound
> Of dulcet symphonies and voices sweet.
> (1.710–12)

Milton compares the completion of this monument, whose construction he likens to the bellowing of a pipe organ, with an exhalation to depict its builders as meteorological beings and to anticipate other moments in the epic when Satan will channel his breath into creating charming sounds. Exhalation comes into play, for instance, when Satan simulates voices with the "organs" of Eve's imaginative faculties and the serpent's "[o]rganic" tongue (4.802 and 9.530). At these junctures, the narrator cites "inspir[ation]" (inblown air) as a possible agent of the implanted fraud (4.804 and 9.189). The interallusiveness of these three episodes underscores the role that Satan's meteorological being plays in conducting sin into the world. The bodies of Eve and the snake are activated by exhalation, just as the "pipes" of Pandaemonium fill up from "one blast of wind" (1.708–9). And, like their infernal counterpart, both Eve and the serpent serve as instruments for generating deceptive sounds: Eve's mind produces the "gentle voice" in her dream that she wrongly thinks belongs to Adam, and the serpent vocalizes Satan's false arguments for eating the forbidden fruit (5.36–37).

God permits the "midnight" or "black" vapors in which Satan steals into Paradise and enters the snake to remain on earth after the Fall. He sanctions their presence by imposing similar climate changes in book 10. Spoiling not only the weather, but also the earth's acoustics, "black air" persists in Eden and terrorizes the sin-wracked Adam, loudly echoing his lament and clouding his judgment (9.159, 9.180, and 10.845–62). The precise method by which Satan instrumentalizes the air to corrupt and vocalize living creatures is the subject of the following chapter. Since fraudulent or dissembled sounds are a direct symptom of climate corruption, we will continue to examine Milton's depiction of the demonic embodiment and appropriation of meteorological processes even as the focus returns to sound and speech.

CHAPTER 4

"How Cam'st Thou Speakable of Mute"
Satanic Acoustics in Paradise Lost

In his 1936 essay, "A Note on the Verse of John Milton," T. S. Eliot claims that Milton's blindness "helped him to concentrate on what he could do best." This was, in Eliot's opinion, Milton's ability to write superbly musical poetry.[1] But for Eliot the genius of his sound is also the sign of his poetic limitation. In his zeal for the aural, Milton neglects the other senses, producing imbalanced poetry in which "the inner meaning is separated from the surface."[2] Eliot's backhanded praise of Milton's "auditory imagination" thus begins by echoing traditional acclaim for the "organ music" of Milton's blank verse before joining in the complaint of F. R. Leavis and Ezra Pound, who equated Milton's "orotundity" with mediocre poetry.[3]

Controversy over Milton's imposing sound effects has abated since Eliot's time. But we can still learn from his contention that Milton's blindness and musical inclination produced poetry that is, above all, acoustically imaginative. Scholars have often noted the play of sounds in Milton's lines—their syntactical arrangement, rhythm, alliteration, repetition, and so on—but comparatively few have sought evidence of Milton's aural imagination in the figuration, characters, and larger narrative structures of *Paradise Lost*, which, along with his style of versification, reflect the poet's distinctive aural concerns. The studies that do cover this terrain tend to look at the political or cultural meanings of Milton's music in early modern England.[4] As Matthew Steggle and Beverley Sherry suggest, the acoustical as well as semantic and musical qualities of Miltonic sounds warrant further critical attention.[5] Milton's wariness of sensuous sound, often attributed to a Puritan bias against polyphony or verbally impoverished forms of music, may be more definitely explained by Milton's metaphysical understanding of the fallen atmosphere and its satanic acoustics.[6]

Of all the notable acoustical features of Milton's epic, satanic aurality stands apart for its centrality in the episode on which the narrative crisis depends. For the Fall to occur the serpent must speak. Before Eve eats the forbidden fruit, her innocence is threatened by the sound of the serpent's speech, which "[i]nto her heart too easy entrance won" (9.734). The penetrating character of this acoustical attack can be attributed to the fact that "sound . . . for Milton . . . is unmistakably corporeal."[7] Satan's identity as an aerial being who has a powerful sway over the atmosphere is certainly also at play. In the previous chapter, I argue that Milton associates Satan with the "prince of the power of the air" mentioned in Ephesians 2:2 and uses this doctrinal point and its traditional elaborations to characterize the fallen angels throughout his epic poetry as powers of air and weather. Milton's depiction of the devils in *Paradise Lost* as elementally similar to, and manipulators of, the atmosphere anticipates their role in *Paradise Regained* as rulers of the postlapsarian middle air.

Despite the basic physical connection between sound and atmosphere, scholars have failed to recognize the codependence in Milton's poetry of acoustical and meteorological representation.[8] The Paradisal airs that carry endlessly mutable praise to the Creator and move tunefully through the garden's leaves illustrate the fundamentally acoustical condition of the atmosphere (5.180–84 and 4.264–66). Satan, the prospective "[p]rince of the air," radically exploits this atmospheric condition (10.185). His success with Eve, and indeed, his whole office in the fallen world as man's deceiver, may be attributed to the cultivation of a studied acoustics that capitalizes on his pneumatic being. I will argue that the mechanical and magical instruments Satan uses to produce deadly sounds prior to the temptation prefigure the method he employs to produce the serpent's voice and that these technologies arise directly from his meteorological agency.

Several insights emerge from a reading of the technical means of production and material basis of satanic sound in *Paradise Lost*. Taking a global view of diabolical acoustics, rather than focusing solely on the devils' fallen music, allows for comparison of diverse aural phenomena—instrumental symphonies, the discharge of cannons, cries of anguish, even musically unaccompanied speech. It also reveals that the conditions of sonic production and the integrity of acoustical material define the potential of sound to corrupt more than, for example, its balance of semantic intelligibility and harmony. I differ therefore with Erin Minear's suggestion that wordless satanic "music proves more powerful than the hollow rhetoric," because it implies that Satan used a less than optimal weapon to tempt Eve, and also misunderstands the technically similar production and identical substance of both satanic music and the serpent's speech.[9] Hell's music and the serpent's rhetoric are made of the same stuff—satanically compromised air—and are manufactured in the same way—instrumentally; they only *seem* different because Satan has switched instruments. Hence, aural contamination may occur in *Paradise Lost* more insidiously than previously supposed at a physical, nonrational level, not simply at the level of words or discernable harmonies, but rather on an elemental scale, where the actual material of sound and its physical disposition are embodied. That satanic sound is simultaneously an artificially created and environmentally integrated phenomenon, moreover, illustrates how Milton's vitalist universe accommodates and adapts aspects of mechanical materialism.

THE MAKING OF SATANIC INSTRUMENTS

The episodes in books 4 and 9 wherein Satan causes Eve to dream and the serpent to speak contain clues as to how Satan uses his meteorological power to produce acoustical deceptions. When Satan is discovered at Eve's ear, "[a]ssaying by his devilish art to reach / The *organs* of her fancy," the narrator uses an acoustical pun to depict one of the possible routes into her mind (4.801–2; emphasis added). Eve's fancy is an organ being played by Satan. Arguing that the "possibility of sin" enters the world through a voice that cannot be assigned singly to Satan, Stephen Hequembourg finds "no ground for asserting that Satan spoke, scripted, or serenaded" when he attempted to suborn Eve.[10] True, the text provides no evidence that Satan transfers verbal matter into Eve's mind, but there are compelling reasons to think that Satan subjects her to noxious sound when he manipulates the organs of her fancy to "forge / Illusions as he list, phantasms and dreams" (4.802–3). Milton places the devil at Eve's ear (what other kind of sensation passes through the ear but sound?), and the forgery he applies to her imaginative "organs" suggests the forge bellows used to power such instruments.[11]

Anticipating the smithing diction of forgery, the verb "assay," which describes Satan's trial of Eve's faculties, also evokes the process of testing the composition or purity of metals.[12] This metallurgical meaning converges with yet another sense of "assay," first recorded in 1665—to sound the depth of something.[13] Given these semantic shadings, Satan's initial mode of *sounding* Eve is investigative; he plumbs her psychic depths as navigators measured the depth of the sea. Furthermore, Satan's reliance on the metallic properties of instruments, as I will demonstrate, also underlies his assaying of Eve's acoustical potential and moral corruptibility. The presence of organ terminology suggests that Satan's sounding of Eve is productive in addition to probing. The emanative and affective nature of his acoustics is evident in the text's other possible explanation of how Satan manipulates Eve: by "inspiring venom" (4.804).

This inbreathed venom aims to taint Eve's animal spirits and inflate her desires (4.808–9). The word "inspire" is used again in book 9 to characterize Satan's transformation of the serpent's "brutal sense" into intelligent faculties (188–89). That Milton in each episode uses this verb, which denotes blowing or breathing into, to depict potential moments of contamination stresses the distinctly aerial conveyances through which Satan accesses and influences God's creatures.

The connection between Satan's meteorological identity and the serpent's acoustical animation is perhaps most manifest when Satan approaches the serpent as "a black mist" and enters "at his mouth" (9.180, 187). But the production of the serpent's speech is not simply a matter of possession. As with Eve's dream, which is engendered either by the organlike action of the fancy *or* by the more direct influx of "inspiring venom," the manufacture of the snake's speech, with "serpent tongue / Organic, *or* impulse of vocal air," hinges on two alternative explanations (9.529–30; emphasis added). In Eve's case, the options given for Satan's operation have been treated as identifying the different inferior faculties susceptible to demonic influence.[14] Taken alongside the description of the serpent's animation, however, the account of the dream's inception fits a pattern of alternation that points to different aspects of satanic instrumentation. By offering these alternatives, Milton implies that Satan has the ability to affect sounds via both the movement of *organs,* the actual apparatuses of sense and speech, and more mysteriously, through the spiritual transference or pulsation of air. Milton's refusal to identify which method—mechanical or spiritual—Satan employs is indicative of a representational strategy used throughout the epic that, as we shall see, does not privilege one or the other side of the rebel angel's being.

Organs had a deep personal significance for Milton. John Aubrey records that he "had an organ in his house; he played on that most."[15] Outside of his home, Milton might have heard the music of John Tomkins, organist at St. Paul's and likely an associate of the elder John Milton, or that of the famous organist Frescobaldi, whom he could have heard while in Rome, mixing in the society of Cardinal Francesco Barberini.[16] Growing up the son of a composer no doubt afforded him ample exposure to the instrument. Milton seems to have taken some interest in the organ's history, noting in his commonplace book when it was first brought to France (*CPW* 1:383).[17] Not surprisingly, then, in *Of Education* the organ is twice recommended as an instrument that should be played after dinner for "recreating" the spirits (*CPW* 2:410).

That the poet grew up around organ music and was himself an organist goes a long way toward explaining why the instrument is a recurring figure in

Paradise Lost; but another rationale lies in the word's multiplicity of meanings, which include musical, biological, and mechanical senses. In Milton's time the word *organ* might denote a pipe, a specific body part, or any kind of mechanical instrument, such as a piece of artillery.[18] All of these senses are at play in Milton's descriptions of the bodily mechanisms through which Satan manipulates Eve and the serpent. The phrases "organs of her fancy" and "serpent tongue / Organic," which clearly designate the bodily sensitive faculties that receive Satan's attempts, also represent Eve and the serpent as Satan's musical instruments and machines. As we shall see, the organ is a leitmotif that connects the musical, mechanical, and bodily apparatuses through which Satan tempts and deceives.

Early in book 1 the fallen angels exhibit their command over both instruments and the atmosphere. Hell's "dusky air" attracts and disburdens them (1.226). It bears Satan's "unusual weight" when he rises out of the flaming lake and it is later pumped through flutes and recorders "blowing martial sounds" (1.227, 540). "Breathing united force with fixèd thought," the angels march silently along to piped music that "charmed / Their painful steps o'er the burnt soil" (1.560–62). They are revived not only by air and music, but also by metals. Instrumental metal described as "[s]onorous" rouses the martial spirits of the angelic hosts, and they lift their glinting weapons on high:

> Ten thousand banners rise into the air
> With orient colors waving: with them rose
> A forest huge of spears: and thronging helms
> Appeared, and serried shields in thick array
> Of depth immeasurable.
>   (1.540, 545–49)

Finally, when they raise their imperial ensign, the "warlike sound / Of trumpets loud and clarions" rings out (1.531–32).

The striking prevalence of metals in this scene and the emphasis on their contact with air is intriguing. Why does Milton go to such lengths to enumerate the devils' weapons and surround them with sonorous music, breath, and the billowing wind overhead? The musical breezes that surround and permeate the defeated rebel angels portend their future status as rulers of the postlapsarian air. But the question of the metals remains. Aristotle attributes the generation of metals to the submersion in the ground and the condensation of certain vapors that he calls exhalations. He conjectures that moist exhalation turns to metal through cooling and by coming into contact with rocks.[19] By Aristotle's logic, then, metal should emerge from the devils' contact with hell's rocky surface, since, as spirits of air, the fallen angels resemble exhalations trapped beneath the ground.[20] They are likened to dense vapor when Satan, summoning them from off the burning lake and onto "firm brimstone," is compared with Moses calling up the locusts over Egypt in a "pitchy cloud" (1.340, 350). The demons are surrounded by metal instruments as a consequence of their moving like a front of vaporous air through hell's atmosphere.

The comparison of the devils' ensign with "a meteor streaming to the wind" prefigures their meteorological potential and plans for atmospheric domination (1.537). The image puns on the definition of the word μετέωρον, of which "meteor" is a transliteration: "something raised up."[21] The account of the streaming ensign thus symbolically corresponds to the defiant hell-rending shout "upsent" by the devils and the rising motion of their weapons (1.541). In the Renaissance the term *meteor* "covered all atmospheric processes and anomalies," including the weather itself as well as comets, stones, and metals.[22] The rising motion of exhalation underlies all of these phenomena and characterizes the devil's acoustics.[23] This crucial description of the fallen angels raising their imperial banner elucidates the moment when Satan raises passions in Eve, "[b]lown up with high conceits" from the vapors of her animal spirits (4.809; emphasis mine). Meteorlike, these prideful notions are the remnant or "flag" that Satan leaves behind in her mind. By mixing with her spirits, Satan's venomous inspiration establishes the internal physio-meteorological climate that renders her a fit instrument of his purpose.

The potential movement of satanic breath through the veins of Eve's body parallels the action of "liquid fire" and molten ore through the venous matrix of Pandaemonium (1.701). The conceits used to describe how the devils construct their great capitol by preparing and recasting liquid gold imply their material presence in the structure of Pandaemonium, which as they build they infuse with their spirituous being:

> As in an organ from one blast of wind
> To many a row of pipes the sound-board breathes.
> Anon out of the earth a fabric huge
> Rose like an exhalation.
>
> (1.708–11)

The first simile links the fallen spirits with organ-blowers who pump the "blast of wind" through the organlike mold. But as flexible powers of air, the spirits are also materially associated with the fluid substance that passes like wind through its pipes and hardens into the golden walls of the palace. The second simile solidifies this connection. The meteorological image of "exhalation" explicitly identifies the demons with their edifice. The golden, metallic fabric of Pandaemonium is *like an exhalation* because it was made by exhalation-like beings and is affiliated with their substance.

By implying a physical relationship between air, gold, and the pipe organ's music, Milton elaborates an ancient theory that attributed the sonority of metals to their porous and aerated internal structures.[24] Albertus Magnus believed that gold, silver, and copper are more resonant than other metals because they contain a superior balance of "subtle water and subtle earth," and also, a substantial amount of vapor.[25] Albertus Magnus writes, "[f]or this reason these metals are strongly resonant and retain the sound for a long time, because they are full of air, and, when vibrating as a result of a strong blow, they continuously expel air from themselves."[26] By alluding to the airy, exhalation-like properties of the gold used to construct Pandaemonium, Milton underscores its function as an acoustical space, productive of sweet music and the charming sound of the devils' political rhetoric.

That pipes, recorders, trumpets, and clarions are intimately connected with the demonic in *Paradise Lost* is not wholly surprising; according to an ancient tradition, wind instruments were considered less noble than strings and were thought to promote the passions.[27] Yet wind instruments also appear in heaven's symphonies, so Milton's treatment of pipes and horns appears to complicate this classical prejudice.[28] Milton's use of the pipe organ—his favorite instrument—to characterize the construction of Pandaemonium where Satan and his followers plot the Fall is similarly perplexing. His closeness to the instrument may have uniquely positioned him to appreciate what might be characterized as its central subterfuge, that a single operator may, with relative ease, create and control a massive, almost unearthly sound. While susceptible to the sublime power of the organ's sound, the organist is always perfectly aware of its cause: an elaborate network of bellows, tanks, stops, and pipes that transform and amplify the machine's initial source of air. Because the organ's miraculous-seeming sound is actually highly engineered—Marin Mersenne called the instrument "one of the most admirable pneumatic machines ever invented"—it aptly symbolizes the artifice of satanic acoustics.[29] The organ simile implies, then, that Satan achieves his impressive-sounding transformation of hell's soil through mere artifice or workmanship. This may have aroused disdain from some of Milton's early readers, for whom the study of mechanics would not have qualified as a liberal art. When John Wilkins published *Mathematical Magick* in 1648, he lamented the persistent bias against practical or artificial (as opposed to divine or natural) investigations in philosophy and declared they should be treated "with greater industry and respect, than they commonly meet with in these times."[30] As late as the nineteenth century, Leigh Hunt hesitates to call the organ a "machine," although a more suitable term than "instrument," because of the lasting stigma: machinery brings "the mechanism itself, however fine and skillful, somewhat too strongly before us."[31] Milton's simile *deliberately* brings the mechanism before us, inviting us to inspect the organ's sound-board and many rows of pipes and imagine the passages, like those of the building's "various mould," that carry air to each row (1.706). While contemplating this fascinating description involves a great degree of pleasure, the association of fallen angels with organ builders, nevertheless, reflects their demoted metaphysical status. That Milton has Mammon, "the least erected spirit that fell," lead the excavations for the building project expresses the low standing of the mechanical arts in hell (1.679).

In addition to disparaging the devils' accomplishment, the analogy of Pandaemonium to a mechanical pipe organ associates their work with magic. Wilkins references the commonplace confusion of mechanical operation with magic when he writes of his book's title: "This whole Discourse I call *Mathematical Magick*, because the art of such Mechanical inventions as are here chiefly insisted on, hath been formerly so styled; and in allusion to vulgar opinion, which doth commonly attribute all such strange operations unto the power of Magick."[32] His point is that geometry or applied mathematics, rather than magic, lies behind the marvels that philosophers of old veiled under "mystical expressions, as might excite the peoples wonder and reverence, fearing lest a more easie and familiar discovery, might expose them to contempt."[33] Magic, whether genuine or purported, is employed in the construction of Pandaemonium. Its indoor lamps are suspended by "subtle magic" and its foundations are laid with "wondrous art" and "strange conveyance" (1.703, 707, 727). Such language obscures the devils' actual means of accomplishing their engineering feats, just as the ancient philosophers in Wilkins's account veiled their arts in secrecy. But the organ simile, by openly depicting the instrument's mechanism, discloses the constructedness of hell's temple.

Milton's use of the alternate strategies of veiling and revealing to describe the devils' assembly of their capitol building is calculated. By integrating simile with direct representation and layering allusions to magic with detailed accounts of the demons' engineering techniques, he keeps the precise nature of satanic industry ambiguous. The devils' powers are magical in that they utterly excel the industry of human beings, and yet their reliance on artifice or craft reminds us that their abilities are less than divine and undeserving of admiration.[34]

If the entanglement of magic and mechanism in Milton's description of Pandaemonium betrays something like Wilkins's skepticism of the occult, it also paints an unflattering picture of the acoustical machines that interest Wilkins and many of his contemporaries. Milton's idea that exposure to natural forces may automatically activate musical instruments is neither original nor purely fictional for his time. The magnificently illustrated *Mechanica Hydraulico-Pneumatica* (1657) and four-volume *Magia Universalis Naturae et Artis* (1657–59) by Gaspar Schott (1608–1666) as well as the influential *Musurgia Universalis* (1650) by Schott's teacher, the Jesuit polymath Athanasius Kircher (1601–1680), offer full of accounts of musical automata that are activated by environmental forces, for example, the hydraulic organ, which uses a water source to produce wind music.[35] In describing these devices, the texts assume a disinterested tone, providing intricate detail and extensive diagrams showing the arrangement of the tanks and pinned cylinders within the machines. The transparency of Schott's and Kircher's texts as to the mechanical workings of these organs is comparable to that of Milton's description, which is forthcoming as to the construction of hell's instrument. However the books' celebration of the organs fits into a broader philosophical program that claims the occult effects of music and portrays magic as a means of harnessing unseen connections within the universe.[36] Schott and Kircher separate the spirit world from the aforesaid instruments and profess to deal merely with natural magic, but their Catholic perspective and authorship would have alarmed a reader like Milton and cast marvels like the apparently self-playing organs pictured in figures 1 and 2, as suspect.

Wilkins also mentions an automatic virginal that, much like Satan's instruments, generates sound through its clever engineering and the application of meteorological power. This device, attributed to the inventor Cornelius Drebbel (1572–1633), allegedly played music when placed in the sun and would cease to play when removed from sunlight. "The warmth of the sun," Wilkins explains, "working upon some moisture within it, and rarifying the inward air unto so great an extension, that it must needs seek for a vent or issue, did thereby give several motions unto the instrument."[37] Wilkins's empirically exact account of this remarkable instrument explains how a simple movement of air, or pneumatics, can propel a mechanism that produces what seems a magical, self-generating sound. By allowing readers to "see"—at least partially—both the mechanism of hell's organ and the airy spirits that pneumatically bring it to life, Milton similarly exposes Pandaemonium's apparently magical acoustics as the product of artful engineering. Milton and Wilkins part ways,

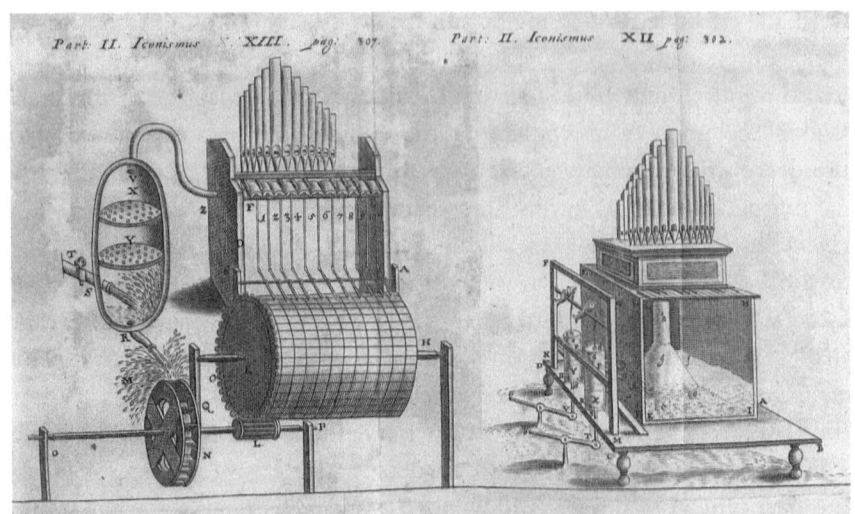

Figure 1. Engraving of an automatic hydraulic organ (*left*) and a traditional Vitruvian hydraulic organ (*right*), in Gaspar Schott, *Magia Universalis Naturæ et Artis*, vol. 2 (Herbipoli [Würzburg], 1657–58), 306–7. (RB 476785, The Huntington Library, San Marino, California; photo courtesy of The Huntington Library)

however, in their regard for mechanical genius; for Wilkins it opens up wonderful possibilities,[38] whereas in Milton's description of hell, it is the devil's handiwork.

We have seen from Aristotle's theory of metals, which remained part of the meteorological tradition through the Renaissance, that the aerial bodies of Milton's devils are implicated in the very material of their instruments. Likewise, the natural philosophy of Albertus Magnus—which suggests that the more air contained in the metal instrument, the better the sound—implies that the quality of satanic acoustics depends on their direct involvement in the fabrication of their instruments. This explains in part why wind and metal instruments that appear in heaven become weapons in the hands of the devils. The narrator does not describe the creation or mechanical workings of heaven's pipes, organs, or golden harps. They appear complete, "ever tuned," and organism-like in their labor to please God with song (3.366 and 7.594–97). In contrast, the poem emphasizes the making and mechanisms of the instruments of hell, thereby disclosing the corporeal yet perversely artificial relation between the demons and their sound makers.

The corporeality of satanic instrumentation in *Paradise Lost* may also evoke and satirize Catholic ceremonies meant to protect consecrated metal from demonic spirits. The Catholic ceremony of "baptizing" church bells, a practice

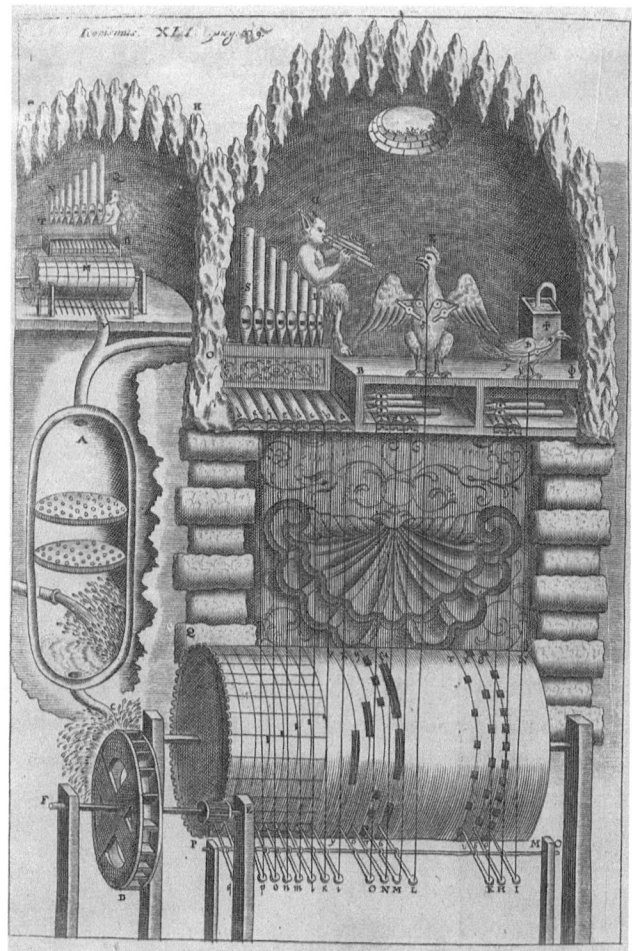

Figure 2. Engraving of hydraulic automatic organs with mechanized bird ornaments and a statue of Pan in a (possibly subterranean) grotto, in Gaspar Schott, *Mechanica Hydraulico-Pneumatica* [. . .] (Francofurtensi [Frankfurt], 1657), 415, plate xli. (RB 709561, The Huntington Library, San Marino, California; photo courtesy of The Huntington Library)

that dates at least to the eighth century, was thought to imbue them with sounds capable of repelling Satan and his agents of weather.[39] According to the service for the benediction of bells in the 1595 Roman pontifical of Clement VIII, after bathing and anointing the bells with the sign of the cross, the bishop would give the following blessing: "And when its melody shall fall upon the ears of the people, may they receive an increase of Faith; may all the snares of the enemy, the crash of hail-storms, hurricanes, the violence of tempests be driven far away; may the deadly thunder be weakened, may the winds become salubrious, and be kept in check; may the right hand of Thy strength overcome the powers of the air, so that hearing this bell they may tremble and flee before the standard of the holy cross of Thy Son depicted upon it."[40] Medieval and early modern church bells across Europe bore inscriptions that echo this challenge to the "powers of air."[41]

Verses from the late fifteenth-century bell tower at Gulval Church, Cornwall, for instance, declared that its pealing bells had the power to banish whatever forces—banal, meteorological, or demoniacal—might impede one from attending church:

> Who hears the bell, appears betime,
> And in his seat against we chime.
> Therefore I'd have you not to vapour,
> Nor blame ye lads that use the Clapper,
> By which are scared the fiends of hell,
> And all by virtue of a bell.[42]

Belief in the efficacy of church bells encouraged many parishes to ring them during thunderstorms.[43] Well into the nineteenth century, Longfellow dramatized this custom in the prologue to *The Golden Legend,* which depicts the bells of Strasbourg Cathedral defending the church against Lucifer who attacks in the form of a storm. His thwarted "Powers of the Air" complain,

> All thy thunders
> Here are harmless!
> For these bells have been anointed,
> And baptized with holy water!
> They defy our utmost power.[44]

The structure of Pandaemonium resembles these parish bells materially and symbolically. Made with the resonant material of gold and compared with an exhaling pipe organ, the "archèd roof" of Satan's temple is designed to reverberate like a bell (1.726). The freshly christened appearance of Satan's temple, "[n]ew rubbed with balm," corresponds with the custom of applying holy water and oil to the consecrated bells (1.774).[45] Similarly, the palace's pneumatic origins, in a hill that emanated "fire and rolling smoke," recall the ceremonial practice of fumigating *campanae* with incense (1.671).[46] Like the "thick" and "airy crowd" of spirits who fill the palace as if they are fairies making "jocund music," fumes of incense flow into the bell during *its* inaugural ceremony, and invest it with talismanic, acoustical power (1.775, 787).

And yet, the moral function of the bells of Christendom is diametrically opposed to that of Pandaemonium, however structurally similar to them Satan's palace may be. The acoustical properties of Satan's temple consolidate and reinforce his power over the air, rather than dispersing it as the bells were supposed to do. Acting as a literal sounding board for the devils' machinations, the acoustics of Pandaemonium invert the apotropaic effects ascribed to pealing church bells or similar metal instruments like the ancient Roman *tintinnabulum* (demon-repelling wind chimes).[47] This key difference gives Milton's invention satirical energy. Without denying that "sonorous metal" can influence atmospheric phenomena, Milton uses Pandemonium to critique those who employ musical instruments to subdue or control the spirit world.

Besides the Catholic clergy, other potential targets of this satire include Paracelsian alchemists like Heinrich Khunrath (1560–1605) whose major treatise *Amphitheatrum Sapientiae Aeternae* was published posthumously in 1609. Penelope Gouk has argued that Khunrath and his circle regarded instrumental music as essential to alchemical practice because it enabled the philosopher to improve his spiritual health and more successfully commune with God.[48] "Particular musical instruments were thought to provide access to the human spiritus through their harmony, a process that made it possible to restore balance between body and soul, and especially to alleviate melancholy."[49] Not only can music refresh the alchemist, whose soul was particularly susceptible to melancholia, it also protects him from evil spirits.[50] The enthusiasm for instrumental music that obtained in Paracelsian alchemical circles around 1600 sheds light on the Miltonic phrase "sounding alchemy" (2.517). This obscure phrase is usually interpreted as a synecdoche for hell's trumpets, but it also likely alludes to the mystical belief that music and alchemy are mutually necessary for achieving good health and spiritual understanding.[51] Milton's usage does not reflect positively on the aims or efficacy of this type of alchemy. Surrounded by rousing music issuing from alchemically and meteorologically forged instruments, Satan is analogized to the magus who uses music to rejuvenate his melancholic soul and increase his spiritual prowess. Because hell's music emboldens rather than impedes Satan and his followers, Milton implies the hubris and futility of attempting to ward off evil with musical instruments. This is not to say that Milton denied music's rejuvenating properties; after all, he recommends the prophylactic use of organ music in *Of Education*. More disturbing than a humanistic practice that incorporates musical performance, however, is an alchemical one because of its dubious association with magic and metals.

Part demonology, part satire, the description of the demons' acoustical activities in book 1 of *Paradise Lost* reveals vital information about the satanic production of sounds, often through ironic allusion to manmade acoustics. We are shown that the devils' meteorological power and mechanical skill are physically realized in their instruments, while subtly reminded that there are men who also use instruments devilishly, with spiritual pretensions or technological hubris. Yet, more than simply parodying those who make idols of their instruments, the passages from book 1 also shed light on the ontology of satanic acoustics. Readers learn that satanic sounds are instrumental—they pass through and are transfigured by some kind of device—and that metal instruments in particular, whose very fabric and resonating cavity were each thought to contain air, are prone to satanic appropriation. The devils make the counterfeiting of sweet, sublime, and finally, human sounds into a mainstay of their ongoing rebellion. Instruments or instrument-like mechanisms enable the demons to convert meteorological resources into deceptive sounds like the "dulcet symphonies and voices sweet" that waft out of the pipe-organ-like structure of Pandaemonium and, as we shall see, the words that mysteriously emerge from the serpent's "organic" tongue (1.712).

THE MECHANICAL ACOUSTICS OF DEMONIC BODIES

Much of the writing on instruments in the seventeenth century did not engage with or endorse the possibility of spiritual participation in instrumental sound.[52] Milton's consideration of the demonological implications of wind-powered instruments stands in contrast to, for instance, their representation by the leading authority on instruments in his day, the French experimentalist and scholar of music Marin Mersenne. But Milton's characterizations of satanic acoustics often echo Mersenne's scientific observations on the interaction between the instrument and the air. The main difference between their accounts lies in Milton's allowance that the movement of the wind within instruments may be magically or spiritually empowered.

Mersenne's *Harmonie universelle* (1636) exceeded any study ever before published in documenting musical instruments—their sounds, appearance, specifications, range, and construction. The books on instruments in the *Harmonie universelle* are illustrated with detailed woodcuts representing a wide array of examples from European courts, but also from homelier traditions and non-Western places. Mersenne's work quickly became known in England, partially through the writings of Descartes and Gassendi, and his acoustic propositions now known as "Mersenne's Laws" were rapidly accepted possibly because they upheld an idea shared by many adherents of the mechanical philosophy, that "the universe was constructed according to harmonic principles expressible through mathematical laws."[53] Mersenne's explanation of musical consonance, for example, referred to today as the "coincidence theory" of consonance, proves that the perception of pleasurable musical sounds has a mathematical basis that can be verified by empirical observation.[54] Just as Mersenne's elegant mathematical descriptions of vibrating strings were the fruits of extensive experiments, his descriptions of instruments display an impressive depth of firsthand knowledge. Mersenne's books on wind instruments and the organ are highly relevant to our discussion of satanic instrumentation in *Paradise Lost*.

Mersenne begins his analysis of wind instruments with a definition of wind, which, he says, is "nothing but movement of the air."[55] He stresses that the moving air within musical instruments and the ambient air found in the atmosphere are interchangeable. The winds that operate flutes and organs, in other words, are the same winds that swirl around us, cause storms, and dive underground. This identity makes it possible to rationalize apparently numinous voices of Nature as the soundings of either naturally occurring or artificially contrived instruments: "There could be related the different caverns, pits, and other subterranean places which sometimes produce harmonious sounds, and at other times confused, horrible, and dreadful noises, to this sort of instruments."[56] The famed acoustical marvels of the ancient world, singing bronze birds, noise-making statues, and the oracles are "done by means of pipes, canals and winds."[57] In fact, every anthropomorphic sound we *think* we hear when no one is there ("the cry of the baby or the song of bird, the concert of viols, the noise of tambourines") is due to the fact that the winds imitate instruments as they strike landscapes and structures ("rocks, walls, window holes").[58] Mersenne is a thorough rationalist who wishes to draw back the curtain on prodigious sounds that seem to be the work of supernatural or magical agency by attributing them instead to mechanism: the interaction of matter and movement. But Milton's motives in portraying fallen acoustics are distinctly theological and mythologizing. While he integrates mechanistic explanation into his accounts of acoustics, he also allows sound to be generated and affected by spiritual agents, as we saw in the chapter on *Comus*. Each of the parts that are requisite for the mechanistic production of sound in Mersenne's view—wind and instrument—are also present in Milton's acoustics. But Milton permits that either one or both of these elements may be demonically possessed, whereas Mersenne bars such influences from entering his descriptions altogether. The wind is always just the wind for Mersenne, but for Milton it is sometimes a spirit.

The physical similarity between accidental instruments found in Nature and those of artificial construction enables Mersenne to dispel the idea that sound has occult properties, but it serves the opposite purpose for Milton. That the devils make and play instruments in hell haunts representations of accidental, naturally occurring sound effects, so that descriptions of the howling wind cannot but be read as the expression of demonic being. The rocks and crevices that catch the wind in Mersenne's explanation are analogized in *Paradise Lost* to the cavities of Pandaemonium's "organ" that resound with the demons' voices. After Mammon says his piece at the great council,

> such a murmur filled
> The assembly, as when hollow rocks retain
> The sound of blustering winds, which all night long
> Had roused the sea, now with hoarse cadence lull
> Seafaring men o'erwatched, whose bark by chance
> Or pinnace anchors in a craggy bay
> After the tempest
>
> (2.284–90)

The cavernous structures in this simile, the "hollow rocks" and "craggy bay," repercuss the "blustering winds" of the tempest, just as Mersenne says "different caverns, pits, and other subterranean places" can sometimes behave like wind instruments. Yet all of the details in Milton's vignette—the rocks, the bay, the "hoarse cadence," the winds and the tempest—do more than simply represent the latent acoustical capacity of nature. They also allude to aspects of demonic acoustics. The simile has a dual implication: the applauding devils embody the sounding winds (the "hoarse cadence") inside instrument-like formations such as Pandaemonium, and they also act like instruments themselves ("hollow rocks") retaining and now expelling air at the appointed time. With equal facility the demons can cause winds to bluster, or temper and dissimulate the sound of wind with instrumentation.

The nautical imagery and allusions to drowsiness in the passage recall the liquid quality of Circe's music, which, as we saw in chapter 2, disposes listeners to relax and even to lose their senses. The seafaring men that the simile depicts unwisely find rest in the repercussions of the waves, foreshadowing Eve's yielding to the serpent's words by "much marveling" at the sound of his voice (9.551). Since the echoes of the storm and the animal's speech come from familiar environmental sources, they seem harmless enough. This conceit from book 2 thus prefigures how all aspects of Earth's fallen environment—not only its winds and vapors, but also its geological formations, such as the underground spring through which Satan reenters Paradise—may be implicated in or transformed into satanic instruments, transmitting the devil's malice without arousing excessive fear or suspicion (9.71–73). The earth's "demoniac holds," which *Paradise Regained* foretells are to be purged by Christ's voice (4.628), receive their possessors in *Paradise Lost.*

Mersenne's scientific discussion of the wind used to sound instruments also anticipates Milton's representation of the devils as makers and users of instruments. Mersenne's view that the air is full of vapor generated by the environment and by organisms—and that this humid air may serve as the wind within instruments, sounding and saturating them all at once—underpins a key aspect of satanic acoustics in *Paradise Lost.* He claims "it is certain that the instruments (of which we are speaking) are able to sound with every sort of wind, whether it comes in as the simple motion of air, as that of the bellows which serve organs and musettes, or whether it is mixed of vapors and water, like that of the mouth, which is so full of moisture that the reeds and tubes of the instrument become all damp and wet."[59] According to Mersenne, the musician's breath does not merely pass through the instrument to make it sound, but rather deposits itself in its parts and mingles with them ("the reeds and tubes of the instrument become all damp and wet"). A similar physical assumption underlies the Miltonic idea that demonic winds merge with and morph into instrumental formations. Mersenne's image of moisture from the musician's body being transferred into and absorbed by the instrument recalls the confusion of the fallen angels' breath with their martial instruments as well as the integration of their exhalation-like bodies with the animating wind inside Pandaemonium.

Mechanical realism similarly motivates Milton's depiction of the interaction and integration of the devils' vaporous bodies with metal instruments in heaven. Raphael's description in book 6 of the rebel angels being crushed by heaven's uprooted mountains plainly illustrates how their substance may be transformed into the very medium of instrumental sound. As noted in the previous chapter, the pun on the word "wind" in Raphael's remark that Satan's troops took a long time to *"wind /* Out of" their suits of armor compares the crushed spirits to the meteorological forces they will control in the fallen world (6.659–60; emphasis mine). The play-on-words also suggests that the spirits' resemble the "wind" inside of instruments.[60] As in hell their exhalation-like bodies supply the wind and music emanating from its grand pipe organ, so in heaven do they expel sounds from pipelike chambers:

> Their armour helped their harm, crushed in and bruised
> Into their substance pent, which wrought them pain
> Implacable, and many a dolorous groan.
>
> (6.656–58)

The word "pent" used to describe the rebel angels' confinement within their metal armor evokes Francis Bacon's characterization of the disposition of the air inside of wind instruments and pipe organs. Bacon uses this term repeatedly to describe the necessary physical conditions for propagating sound: "where the air is pent and straitened, there breath or other blowing, (which carry but a gentle percussion) suffice to create sound; as in pipes and wind-instruments."[61] The acoustical diction Milton uses to describe the constriction or penning in of the angels' vaporous bodies analogizes them to the air within musical instruments and represents their groans as the notes emitted by flutes or a pipe organ.

Early readers would be especially apt to visualize the demons as embodying instruments because of familiar representations in the visual arts. Two artworks inspired by Athanasius's *The Life of Antony*, for instance, depict demons playing or carrying instruments and with pipe or horn appendages.[62] Jacques Callot's 1635 etching, *The Temptation of Saint Antony*, is particularly creative and grotesque in the ways it imagines devils using their bodies instrumentally (see fig. 3). One demon strums the lute and another plays a horn protruding from his anus. In the foreground, a large scaly creature—wheeled and fired like a cannon—blasts smoke, ammunition, and presumably a terrible noise into the fray. Another devil, from the prospect of a cloud, spews out toxic sounds and gasses from his buttocks. Callot's image thus graphically captures the early modern conception of the demonic body. Its organs, artificial rather than biological, are incorporated musical instruments, engines, and implements that enable the demon to harass man. Satan risks becoming like these hybrid creatures as he increasingly embodies his status as prince of the air and uses it as a platform for acoustical warfare. In the mountain-throwing episode, his troops fall victim to their own violent acoustical methods when crushed by their armor and unwillingly transformed into instruments. Their pitiful concert of groans darkly doubles hell's "dulcet symphonies." It also anticipates Satan's instrumentation of Eve and the serpent, and finally, the moment in book 10 when the devils are reduced to hissing serpents. Trapped in their suits of armor, their sounds are involuntary—a condition of their bodily imprisonment and punishment.[63]

Raphael's account of the war in heaven reveals the rebel angels' meteorological procedure for lacing sounds with fraud. Satan's "hollowed" engines—"deep-throated" instruments with "mouths" that roar, belch, and exhale smoke—illustrate the duplicitous and destructive aspects of his acoustics (6.574, 576, 586). The artillery causes havoc for the good angels in the usual way, by scattering their ranks with "balls / Of missive ruin" (6.518–19). But it also serves up

Figure 3. A band of demons seated above the beleaguered saint provide musical accompaniment in this detail from the etching by Jacques Callot, *The Temptation of Saint Antony,* 1635. (Photo courtesy of the William M. Ladd Collection, Minneapolis Institute of Art; gift of Herschel V. Jones, 1916)

a different kind of ammunition in the form of a piercingly loud and deceptive blast. By siphoning "sulphurous and nitrous" meteorological materials into their cannons, Satan and his crew attempt to make a weapon that sounds as terrifying as thunder, the coveted armament of God (6.512).[64] They succeed at least in producing an exceedingly violent and startling sound, "[e]mbowell[ing] with outrageous noise the air" (6.588). In one respect, this language creates the impression that ethereal air is corporeal and endowed with viscera vulnerable to the mangling force of demoniac sound. But the word "embowel" may also suggest the animal bladders used as the bagpipe's wind reservoir, implying that the heavenly air undergoes a kind of violence associated with satanic instrumentation.[65] As heaven's "materials dark and crude" become the explosive charges for the rebels' outrageous artillery, so the air is crudely "convey[ed] into the bowels" of the instrument the devils aim to make of the entire atmosphere: a resonator of false thunder (6.478).[66] The opposing senses of the word embowel, which can denote both the loss of and the filling up of guts, paradoxically render the air an eviscerated yet acoustically repurposed space.

Milton's association of Satan's guns with instruments and emphasis on their booming sounds is not a historically anomalous treatment of cannon fire. Along with the vibrational sounds of stringed musical instruments, the report of guns and artillery was a central subject of acoustical investigation during the early seventeenth century. In *Sylva Sylvarum,* for instance, Bacon refers to the "noise of great ordnance," fired at long distances, as an example of how sound does not immediately reach our ears and travels at a slower speed than light.[67] Several of Bacon's seventeenth-century successors, including Marin Mersenne, Pierre Gassendi, and members of the Florentine Accademia del Cimento, actually attempted to measure the speed of sound using a technique called "blast-timing" which involved "timing the interval between seeing the flash and hearing the report of guns fired at a known distance."[68] Guns were valued in Milton's day not only for their military function, but also their extraordinary sonority.

If the blast of Satan's engines is meant to simulate the intimidating sound of thunder, then the initial appearance of the cannons augurs an altogether different kind of sound. To the "amused" faithful angels, their shape expresses vocality:

> their mouths
> With hideous orifice gaped on us wide,
> Portending hollow truce.
>
> (6.576–78, 581)

As the novel appearance of the cannons momentarily diverts the heavenly soldiers, early modern readers may have connected their arresting appearance with that of the basilisk, an imposing-looking medieval cannon aptly named after the mythical serpent that kills its prey with a glance. Thus, while Adam and Eve could not have intuited the serpentine associations of the weapons in Raphael's tale, early audiences possibly understood the cannons as a direct type of the serpent in the garden, attributing their disarming appearance, shrewd mechanism, and violent acoustics to the latter satanic instrument.

THE SATANIC ORGAN OF SPEECH

Each of the episodes discussed thus far shows Satan making and using different kinds of organs. The grand organ blast of the infernal council, the demons' embodiment of instruments in their metal garments of war, and their fabrication of military engines that are as loud as they are brutally violent, illustrate the rich interplay between the musical, corporeal, and technological senses of organ. The demons' ability to fashion these organs and make each of them sound depends on their meteorological bodies, that is, their similarity to wind and the pneumatic ingredient in metals. Satan clearly draws on his embodied experience with organic devices when he inspires Eve's dream and animates the snake. But how does he counterfeit human speech in the body of an animal?

Satan's acoustical career culminates in his vocalization of the serpent, which under his influence becomes yet another sort of instrument. For early modern readers of *Paradise Lost*, this transformation would have required no stretch of the imagination. As early as the sixteenth century, a wind instrument called the serpent was being used in France to accompany church choirs.[69] This impressive horn—some were over eight feet long—takes its name from the snaky shape of its tube, whose initial "S" curve repeats itself in a wider loop at the bottom of the instrument.[70] That Milton was aware of this instrument is evident from its resemblance to Eden's serpent. Brass or other kinds of metal were sometimes used to make serpents; but they were typically constructed of wood and wrapped with leather.[71] Often, they were fitted with a brass crook, a piece of tube inserted between the mouthpiece and the body of the instrument in order to change its tone.[72] A flash of metal appears on Milton's serpent in the same place. We are told that he has a "burnished neck of verdant gold" (9.501). More significantly, the snake's erect stature as it approaches Eve

> on his rear,
> Circular base of rising folds, that towered
> Fold above fold a surging maze
> (9.497–99)

mirrors the ascending, folded appearance of the horn. It too has a "[c]ircular base"; the lowermost coil curls around until it almost closes. Furthermore the word "base," in Milton's description, may allude to the low register of the serpent whose deep tones were especially valued for filling out the bass parts in choral music.[73] According to Mersenne's account, the serpent is "capable of supporting twenty very strong voices," and its tone may be easily modulated, "so that it will be suitable to join with the soft voices of chamber music, whose graces and diminutions it imitates."[74] The versatility of the musical serpent matches the subtlety of Milton's snake, who shifts his tone mid-argument in response to Eve (9.664–68). If he had not witnessed a serpent being performed himself, then Milton could have seen striking illustrations of them in Mersenne's *Harmonie universelle* (see fig. 4) and Athanasius Kircher's *Musurgia Universalis*.[75]

Milton's characterization of Satan as an operator and maker of instruments appears all the more strategic in that, in his day, serpents were actual

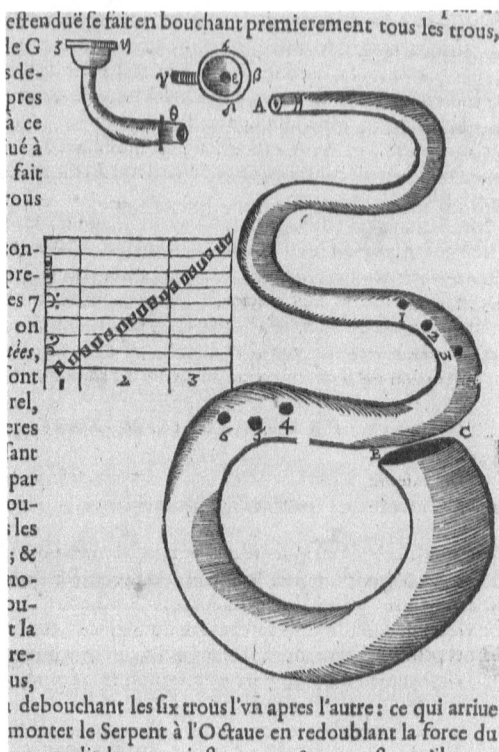

Figure 4. Woodcut of serpent with mouthpiece and crook, in Marin Mersenne, *Harmonie universelle contenant la théorie et la pratique de la musique* [. . .] ([Paris] 1636), 5:279. (Photo courtesy of the Bibliothèque nationale de France)

instruments used in concert music to augment and imitate the human voice. In the late Renaissance, large organs began to be built with pipes that were designed to sound like the human voice.[76] Controlled with an organ stop called the Vox Humana, this feature is still used in theater organs today.[77] The two-pronged account of Satan's method of inspiring the serpent ("with serpent tongue / Organic, or impulse of vocal air") leads Hequembourg to the forked question: "Is the serpent more like a singer or more like a trumpet?"[78] Drawing such a dichotomy on the basis of these phrases, however, misinterprets the word "tongue" and ignores the polysemy of the word "organic" and its allusion to Satan's history of using, making, or embodying instruments. The narrator's interjection after the serpent tries his wiles on Eve, "[s]o glozed the tempter, and his proem *tuned,*" continues the idea that the serpent is being used instrumentally (9.549; emphasis added). Indeed, that instrumental serpents were a feature of the Renaissance musical world suggests that Satan speaks to Eve in the language of instruments—in a "tongue / Organic."

Some pipe organ terminology is necessary to decipher this phrase. Organ pipes come in two different types, flues and reeds; the former kind emits air through a simple slit in the side of the pipe called a "mouth"; the latter type incorporates a metal reed called the "tongue."[79] The *OED* gives 1551 as the earliest date "tongue" was used in this precise context.[80] The positioning of the tongue is essential for the tuning of the organ pipe: "the choice of the sounding frequency is basically made by the reed, and the air column must follow."[81] Thus, the text's equivocation about how Satan manages what we might call "the serpent trick" is not about whether the snake is a complicit singer or a passive trumpet. The serpent undoubtedly serves as an instrument. The textual vacillation refers to, rather, the tempter's acoustical technique. We must keep in mind two conceptions of Satan; the first is of a mechanical genius who uses the tuning mechanism of the tongue to determine the frequency of each note. The second conception (foregrounding "impulse of vocal air") is of Satan as a spiritual and aerial being; enclosed in the serpent, possibly even in its windpipe, Satan vibrates with sound, his aerial body serving as the column of moving air within an instrument.[82]

In the early modern period, the word "impulse" might denote the "[f]orce or influence exerted on the mind" by a good or evil spirit.[83] The serpent's artificial language immediately impresses Satan's spiritual influence on Eve's mind. Its impulse of air travels into Eve's ear and then her mind and heart, causing her to "marvel" greatly at the voice and provoking this demand: "Redouble then this miracle, and say, / How cam'st thou speakable of mute" (9.551, 562–63). Just four lines later, Eve redoubles her own speech, still uttered in the imperative mood: "Say, for such wonder claims attention due" (9.566). Her rhythmic repetition of this monosyllabic word, "say / . . . / Say," echoes the serpent's acoustical impulse in her mind. She resonates to his sounds. Keeping these reverberations alive in Eve's organs—indeed, inspiring her to continue them vocally in the highly echoic language she uses to ponder the interdicted fruit—is crucial to Satan's success (9.745–79). Even after he ends his argument, "in her ears the sound / Yet rung of his persuasive words," its aural power over Eve's mind undampened (9.736–37). These lines bear an unsettling likeness to the description of Adam, added to the poem in 1674, listening to Raphael's magnificent account of the world's creation, and before that, of the war in heaven. After Raphael ends his narration, Adam thinks mistakenly that the angel is still speaking, for "in Adam's ear / So charming left his voice" (8.1–2).[84] Raphael's voice seems to dilate like the serpent's speech in the listener's ear, even after the words are uttered, raising doubts about how safely *any* sound can be conveyed in the organs of its auditors.

Such a doubt may also underlie Milton's anxious desire to find a "fit audience" for his poem (7.31). The perils of communication lie not only in human organs, but also in the passage of sound into written and then printed form. Bruce Smith, a scholar keenly attuned to the embodiment of sound in early modernity, notes that "at every step in the process that transforms a manuscript into a printed text, a body of some sort interposes itself between the act of speaking and the act of reading."[85] On its way to publication, Milton's poem passed through the bodies of the amanuenses who heard and transcribed its words, as well as the various, often metal, apparatuses of the printing house, whose components bore the names of body parts.[86] The "evil tongues," then, of book 7 may be interpreted both as a synecdoche of Milton's corrupt political environment and a suitably organic figure for the satanic acoustics of the fallen atmosphere (26). In commenting on the composition of his epic, Milton acknowledges that evil "dissonance" threatens to invade the sound of his song as it passes through bodies and instruments (7.32). What I have been calling satanic acoustics, the artificial animation of organs by corrupt meteorological materials, therefore, is a potential outcome of reading or hearing *Paradise Lost*, an outcome of which the poem is aware. Milton's wish, implicit in the lines he added to the epic in the final year of his life, is that readers too will be recreated—instead of deluded—by its sound as Adam emerges from Raphael's tale "as new waked" and capable of standing firm (8.4).

CHAPTER 5

Milton and the Barometer
Climate Change in Pneumatic Science

The episode of Milton's tour of Europe in 1638 and 1639 that most influenced his thinking in natural philosophy arguably was his meeting with the blind and aged Galileo, who was then confined by the Inquisition to his home near Florence. The visit evidently affected him deeply. In his critique of licensing addressed to parliament a few years after returning to England, Milton proudly recalls having met the astronomer while in Italy. Decades later, he mentions Galileo by name in *Paradise Lost*—a distinction bestowed on no other contemporary.[1] In a now classic work of criticism, Marjorie Nicholson makes the case that traveling to Galileo's doorstep, as it were, sparked a pivotal change in Milton's view of space: "an Italian 'optic glass' first made him conscious of realms of vision and of thought which his youth had never imagined."[2] Allusions to the satellites of Jupiter, newly observed stars, and the earthlike topography of the moon in *Paradise Lost* are all hallmarks of Galilean astronomy (8.148–52, 7.382–84, 5.261–63, and 8.145–48).[3]

Though the vast and daring cosmography of the epic evidently draws on Galileo's observations with the telescope, we should not assume that astronomy is the only source of Galilean influence in the poetry. The philosopher also pioneered mundane physics, a field that, much like astronomy, was utterly transformed by the advent of a new instrument. Milton's visits to Rome and Florence where he met the "Tuscan artist"[4] and his pupils and frequented academies, placed him geographically, socially, and intellectually proximate to Italian experimentalists who were on the brink of inventing the barometer (1.288). With this simple glass tube they would forever transform the face of science: "Discovery of the barometer . . . changed the appearance of physics just as the telescope changed that of astronomy."[5]

Around the time that Milton traveled to Europe, changes were underway in the history of pneumatics, the branch of seventeenth-century physics that dealt with the mechanical properties of air. Milton's journey coincided with the beginning of a series of experiments performed in Italy (later in France, Poland, Germany, and England) that attempted to produce a vacuum and led to stunning revelations about the pressure and elasticity of air.[6] When Milton returned to London, his Continental ties and activities as a private tutor and polemicist kept him in contact with individuals who were part of this midcentury exchange in pneumatics, which culminated in England with the discovery of Boyle's Law of Gases.

The intersections between Milton's career and this transcultural revolution in physics have not been extensively analyzed.[7] Perhaps the story has received relatively scant attention because the shift pertained to the physics of atmosphere, and air has been a neglected subject of Milton's poetry and early modern literature in general. Though the barometer may have afforded a subtler shift in perspective than optical instruments of the day, it had a remarkable and lasting impact on ideas of the natural world. Whereas the telescope illumined the space beyond the world not visible to the naked eye, the barometer demonstrated tendencies in the air that the body cannot readily sense. As this book has argued, Milton's poetry is utterly interested in the physical and spiritual contents of the atmosphere and how it behaves. Pertinent to his belief in demonological powers of air and their interference with meteorological and acoustical phenomena are reports of an instrument that showed fluctuations in the heaviness of the atmosphere.

Just as Aristotelians doubted the telescope's bold new picture of the heavens,[8] early interpreters of barometric devices challenged their unthinkable implications for physics. The force that upholds the column of mercury in the barometric tube and the contents of the enclosed space above the mercury (was it truly empty?) were matters of intense dispute. At stake were two ancient and unresolved philosophical problems: whether a vacuum can exist in nature and whether air has weight in its "natural place." Milton's poetry reflects the renewed interest that barometric investigation brought to conundrums of the "old physics," particularly, the reassessment of the weight of the air. Scholastic cosmology portrayed air as weightless when nested in its natural place, that is, below the sphere of fire and above the heavier elements of water and earth. But early versions of the barometer upended this world picture and replaced the perfectly balanced sphere of air with a weighty mantle, the bottom layers of which are more compacted due to the pressure and elasticity of air. In key respects, this new understanding of atmosphere as revealed by the barometer resembles the fate of the climate in *Paradise Lost* as it sinks under the effects of sin and satanic influences in the air. The collapse of the harmonious equilibrium between the elements at the Fall and the effect of this disruption on the weight of atmospheric air, thus comprise a fictional echo of the impact of the barometer on the conceptual history of air.

What is a barometer? What does it measure? If readers find the answers to these questions elusive, so did many of the pioneers of the seventeenth-century barometer. For several years, the instrument did not even have a name. The object itself was subsumed by a wider conversation about "Torricelli's experiment" (ca. 1644)—a demonstration designed by the Italian scientist Evangelista Torricelli (1608–1647) who is now recognized as the barometer's inventor.[9] The word "barometer"—derived from the Greek *baro* for "weight"—appears to have been coined, long after the initial experiment took place, in England in 1666 (and almost simultaneously in France).[10] To detect the formation of potential storms, meteorologists today monitor barometric data from around the world for "depressions," shorthand for the phrase "areas in which there is a depression of the barometer."[11] Though Torricelli had intended his instrument to monitor atmospheric conditions the accuracy of early barometers was limited by their sensitivity to changes in temperature.[12] The earliest sequence of air pressure readings was taken in Pisa in 1657–58 as part of the world's first meteorological network known as the Rete Medicea.[13] While contemporaries such as Robert Boyle made intermittent recordings, the barometer was not used in England as a tool for systematic weather observation until the close of the seventeenth century.[14]

Torricelli was among the last and most brilliant of Galileo's students. A budding mathematician living in Rome, Torricelli's talents won him an invitation to work as an assistant to Galileo, whom he joined at Arcetri near Florence just months before the philosopher died.[15] When he arrived in October of 1641, he was 32 years old, by the year and almost to the month, exactly the age of John Milton. Although Torricelli knew Galileo only briefly, the elder man's expertise on vacuums and the behavior of fluids centrally influenced Torricelli's invention. Torricelli likely selected mercury as the active ingredient for the experiment at the urging of one of Galileo's students who had transcribed his research notes.[16] Galileo also made a lasting impression on Milton, who met the savant while he was staying in Florence. However, Milton's visit with Galileo was too early, by at least two years, to have overlapped with Torricelli's stay at Arcetri.

A mere two or three months after Torricelli settled in at Arcetri, the ailing philosopher passed away. His views about the physics of vacuum, however, laid the groundwork for Torricelli's investigation and may have shaped the specific character of his experiment. Galileo's primary contribution was his misunderstanding of a seemingly minor mechanical problem. Unlike the vast majority of thinkers of his day who accepted Aristotle's rejection of void, Galileo believed in the existence of vacuum. But he was incorrect in assuming that vacuum exerts an attractive force to which he attributed the effect of suction pumps and siphons.[17] Despite having been cautioned by Giovanni Batista Baliani (1582–1666) of the importance of the weight of the air, Galileo resorts to the so-called force of the vacuum in the *Discorsi e dimostrazioni matematiche* (1638) (*Two New Sciences*) to explain why pumps are incapable of raising water above a certain height.[18] He reasons that the attractive pull of the vacuum cannot suspend the column of water past the column's breaking point.[19]

In December of 1638, around the time of Milton's first visit to Rome and before Torricelli moved to Florence, *Two New Sciences* arrived in the capital. Afterwards, a group of people in Rome, including Gaspar Berti, Rafael Magiotti, and Athanasius Kircher, organized an experiment to create a vacuum and put its force to the test.[20] They fastened a spherical water flask to the outer wall of Berti's house and connected it to a tube that descended along the wall into an open cask of water below. With the flask and tube completely filled and sealed off from the air, the water was then released through a tap at the base of the tube into the cask below.[21] The whole system behaved much like a barometer; the column of water dropped to a certain height leaving a mysterious space in the flask above. Participants in this event and Torricelli's subsequent experiment appear to have been following Galileo's lead. W. E. Knowles Middleton argues that Berti's group constructed the contraption at Galileo's urging.[22]

Implicit in Galileo's explanation of water pumps was a refutation of the Aristotelian plenum and the fiercely defended Scholastic dogma that Nature abhors a vacuum (horror vacui).[23] Thus it was potentially in hopes of confirming the possibility of a vacuum, a concept associated with Democritean atomism and atheism, that early experiments such as Berti's were attempted and hotly debated. However, interest in the potential creation of an artificial vacuum was not the only factor that motivated and shaped the history of the barometer. Speculation that the air has weight, even when freely dispersed in the atmosphere, also drove the design and interpretation of pneumatic experiments, which in turn demonstrated the phenomenon of atmospheric pressure. Before explaining Torricelli's role in this discovery process and its effect on Milton, it will be helpful to examine some common assumptions about the weight of air that preceded the famous Tuscan experiment.

DOES AIR WEIGH IN AIR? THE ANCIENT DEBATE

Prior to Galileo's era, the physics of weight was grounded, with important exceptions, on Aristotelian natural philosophy. In the fourth book of his treatise *De Caelo* (*On the Heavens*), Aristotle outlines a theory of natural place that explains why the four elements, fire, air, water, and earth, gravitate to their respective positions in descending order around the center of the globe. He uses the terms "light" and "heavy" to describe a body's natural tendency to move upwards or downwards, either towards the outer limit of the universe or its center. The element of earth is absolutely heavy because it sinks below everything else, and fire is absolutely light because it rises above everything else.[24] Channeling these Aristotelian principles, Moloch in *Paradise Lost* believes that absolute levity is the natural motion of the fallen angels:

> in our proper motion we ascend
> Up to our native seat: descent and fall
> To us is adverse
>
> (2.75–77)

In Aristotle, the middle elements, water and air, have a more complicated status: "bodies which possess both contraries are heavy and light in another sense, rising to the top of some and sinking beneath others. Such are air and water, neither of which is absolutely light or heavy."[25] Aristotle provides experimental evidence that emphasizes the contrariety of motions belonging to air: "[I]n its own place every body has weight except fire, even air. It is proof of this that an inflated bladder weighs more than an empty one, showing that if anything contains more air than earth and water, it may be lighter than something else in water but heavier in air; for it rises to the top of water, but not of air."[26] This passing reference to the heaviness of the inflated bladder, however, did not settle the question. Two important successors of Aristotle, Ptolemy (ca. 150 AD) and Simplicius (sixth century AD), disagree with Aristotle's view that the element of air has weight. In performing Aristotle's experiment, Ptolemy apparently derives the opposite result: the inflated bladder is lighter, not heavier, than the empty bladder.[27] This finding supports somewhat Ptolemy's view that water and air do *not* have weight when in their natural places, a principle apparently illustrated by the fact that divers, when submerged in water, do not feel the heaviness of the element above their heads.[28] Simplicius also conducts the bladder experiment and finds that the two bladders weigh exactly the same. Although his results are at odds with those of both Aristotle and Ptolemy, Simplicius sides with Ptolemy's theory of weight: "That air should have weight in its own place and that therefore the inflated skin should weigh more would be still more paradoxical. It would mean that air experiences a downward pull from its own place by a natural motion, which, I believe, cannot make sense according to Aristotle's doctrine."[29] Simplicius tactfully rejects Aristotle's claim that air has weight in its place on the basis of Aristotle's own logic. According to Aristotle an element is heavy insofar as it strives downward. Once it reaches its proper region, Simplicius reasons, it no longer strives to be anywhere and thus would seem to have no weight.[30]

Although these conflicting accounts of elemental weight imply disagreement and confusion on the subject, Simplicius's modification of Aristotle's position vis-à-vis Ptolemy offers a compromise that many later thinkers would find acceptable: the elements do not have weight in their natural places. This notion was so enduring that in 1744 the author of a textbook on experimental philosophy expresses an obligation to dispel it: *"All the Parts of a Fluid are heavy at all Times and in any Situation; which it would be unnecessary to mention, since all Matter is heavy, if a great many Persons had not affirm'd, and given out as an Axiom, that Elements do not gravitate in their own Places; as for Example, that Air does not weigh in Air, nor Water in Water, &c."*[31]

In *Two New Sciences,* Galileo puts this axiom in the mouth of Simplicio (not to be confused with the historical person Simplicius) who often typifies the viewpoints of a Peripatetic philosopher. Simplicio's doubts about weighing the air echo features of the Aristotelian doctrine as modified by Ptolemy and Simplicius: belief in the qualities of heavy and light ("What if it is found that air, instead of gravity, has levity?"); the significance of natural place ("It is undoubtedly true that the elements in their own regions are neither heavy nor light"); and denial the air has weight in its place ("I can't understand of that portion of air that appeared to weigh, say four drachms of sand, how or where this can really be said to have that weight *in air*").[32] Advocated by the credulous Simplicio and quickly refuted by his interlocutors, the doctrine of the weightlessness of air in its region epitomizes the outlook of the old physics.

By the 1630s, it was no longer plausible to claim that air "itself is light and it has this in common with fire," as one twelfth-century treatise avers.[33] Galileo's Salviati sums up what would become the standard new account of the weightiness of air: "Positive weight exists in air, and not lightness as some have believed; *that* is perhaps not to be found in any material whatever."[34] So much for the Aristotelian qualities of heavy and light. But what are the implications of weight on whole the region of air? Isaac Beeckman (1588–1637) and Rene Descartes, writing in 1620–21 and 1631 respectively, answer this question in bold and memorable terms. They reason that the layers of air become increasingly compressed the nearer they lie to the earth.³⁵ Descartes compares the air with the flexible material of wool: "the wool which is upon the earth is compressed by all that which is above, even to beyond the clouds; and this makes a great weight."³⁶ Beeckman pursues the analogy of a sponge: "It cannot be denied that the lower part of water or air is more strongly compressed than the upper part, since it is compressed by its own weight, *as would happen to an immense sponge,* its lower part lying on the earth is packed more tightly than the upper part."³⁷ Wool, a sponge, and later, a spring—these were the familiar objects that rendered visible the potential structure of the air, linking local atmospheric effects such as compression to the whole heft of the sky.

The French scholar Jean Pecquet, whose book anticipated the notion often attributed to Robert Boyle of the "elasticity" of air, echoes the language of Beeckman and Descartes, calling the atmosphere "a heap of spongious, or rather wooly material encompassing the Earth-watrish Globe" and explaining that the "elatery" of the air "doth imitate the nature of a Sponge or Wool."³⁸ Although it would be difficult to establish whether Beeckman's sponge analogy was known before Pecquet's book was translated into English in 1653, it is worth noting that Milton uses the figure in 1634 to describe the absorption of Comus's spells into the "spungy ayr."³⁹ Whether or not sponginess in Milton's usage suggests the air's compressibility in addition to its porosity (as discussed in chapter 2), the diction is striking given that contemporaries used the term in a technical sense to underscore the materiality and mechanical behavior of air.

THE EXPERIMENT THAT CHANGED THE SKY

On his journey home from Italy in June of 1639, Milton visited Calvinist Geneva where he recorded in an acquaintance's book of autographs a Horatian epigraph: "I change my sky [*coelum*] but not my mind when I cross the sea."⁴⁰ The line refers to his having faithfully preserved his national and religious identity while traveling to the heart of Catholic territory. He could not have had any inkling at the time of the pivotal Torricellian experiment, which was to occur a few years later in Florence in 1644.⁴¹ Nevertheless, in hindsight, the inscription foreshadows the striking new conception of atmosphere that proceeded from the region Milton recently visited. Not long after he crossed the sea on his return trip to England, the sky permanently changed, at least as it was understood by Europeans. Beeckman and Descartes may have paved the way for this shift in thinking about the atmosphere, but experimental evidence of air pressure precipitated a real change in opinion, and this evidence came in the form of a simple demonstration with a tube of mercury.

In 1644, Torricelli was philosopher and mathematician to the Grand Duke Ferdinand II, a post previously occupied by Galileo. For unknown reasons, Torricelli himself did not conduct the experiment that bears his name, but instead gave the specifications to Vincenzo Viviani (1622–1703) who carried it out. Viviani is the aforementioned student of Galileo's who may have capitalized on his teacher's insight about the breaking point of mercury by including this ingredient in the experiment.[42] Torricelli received credit for the barometer, in part, because he clearly explained and defended the results of the experiment in a letter to his friend Michelangelo Ricci (1619–1692), which became one of the most influential documents of seventeenth-century pneumatics. Extracts from this famous letter soon crossed the desk of Mersenne, who was the central hub of scientific correspondence at the time.[43] Thus within two years, it was heard in France that Torricelli had made an empty space in a vessel and attributed its cause to the weight of the air.[44]

According to Torricelli's description, the prototype of such an instrument consisted of a tubular glass vessel completely filled with mercury so as to exclude all air. The mouth of the tube was covered by a finger and then inverted and submerged in a basin of mercury. When the finger was released, some of the mercury from the tube flowed into the basin, but the column remaining in the tube decreased to "the height of an ell and a quarter and a finger more," which C. Webster takes to mean 29 inches.[45] The fact that the mercury did completely not fall from the tube when a passage was cleared and instead sank slightly to the height were it stayed suspended, resembled the behavior of the pumps discussed by Galileo in *Two New Sciences* and the outcome of the experiment performed at Berti's home. But Torricelli's explanation argued decisively against previous interpretations: "[W]e discussed this force that held up the quicksilver against its natural tendency to fall down. It has been believed until now that it was something inside the vessel . . . but I assert that it is external, and that the force comes from outside."[46] The external force, Torricelli explains, is nothing more than the manifestation of the weight of air at a given altitude: "On the surface of the liquid in the basin presses a height of fifty miles of air."[47] The height of the mercury in the tube equals "the point at which it is in balance with the weight of the external air that is pushing it!"[48]

Torricelli seems to anticipate that this conclusion would be controversial. Galileo's belief that a vacuum exerts an attractive force suggested the possibility that the void-like space within the tube—not the exterior air—caused the elevation of the mercury. Borrowing a metaphor from a neighboring sublunary zone, Torricelli insists, however, that the force comes from above: "*We live submerged at the bottom of an ocean of elementary air,* which is known by incontestable experiments to have weight, and so much weight, that the heaviest part near the surface of the earth weighs about one four-hundredth as much as water."[49] The passage radically displaces human beings from their presumed terrain. We are depicted not as surface dwellers, but people who live "submerged" at the base of an "ocean of elementary air" (*Noi viviamo sommersi nel fondo d'un pelago d'aria elementare*). This revelation redefined humanity's position in the cosmos and transformed conceptions of the diaphanous matter that envelopes us. Torricelli's heavy, liquid-air analogy resonates with Milton's representation of the atmospheric regions of hell, chaos, and the fallen earth, which he frequently portrays in oceanic terms.

Torricelli's comparison between air and water implies that, regardless of the element, the weight of all matter tends towards earth. Proving that even ambient air displays this uniform tendency, Torricelli's demonstration dealt a severe blow to the Scholastic assumption that the four elements have distinct "spheres" to which they each gravitate separately. Coming on the heels of the collapse of the Ptolemaic world system, the cosmological implications of Torricelli's experiment furthered a conception of the universe forged by Tycho Brahe, Galileo, and others, that was far less segmented than previous models—with their emphasis on solid orbs and terrestrial zones—and fundamentally more integrated and fluid. Foreshadowing "Newton's synthesis,"[50] which brought the sublunary and supralunary realms under a single physics, Torricelli's comparison of the atmosphere to an "ocean of elementary air" recalls an earlier sea change in natural philosophy: astronomy's dalliance with the "fluid-heaven theory."[51] Regaining popularity during the early seventeenth century as an alternative to Ptolemy's explanation of planetary motion, this late-medieval theory likened the free movements of the celestial bodies through the heavens to "birds in the air, or fish in the sea."[52] The humble avian-piscine analogy, an axiom of the fluid-heaven theory, anticipates Torricelli's metaphor of the "ocean of elementary air" by conflating the elements of air and water, and suggesting that great swaths of the universe are composed of liquid-like substances. Armed with barometers and a new consciousness of the weight of the air, John Beale, an active participant in the Royal Society, further develops the marine-atmospheric trope: "that we find the Weather and our Bodies more chill, cold, and drooping, when the *Mercury* is lowest, and the Air lightest, besides other causes, I guess, That as Air is to us the breath of life, as water is to Fishes; so, when we are deprived of the usual measure of this our food, 'tis the same to us, as when the water is drawn ebb from Fishes."[53]

Milton shared some of these bold new assumptions about the mundane climate and the universe beyond. On the one hand, *Paradise Lost* invokes traditional features of the world, such as the outer shell and orbs, and on the other hand, disputes their solidity by depicting them in mixed hydro-pneumatic terms. For example, as the Father looks on from above, Satan, at the beginning of book 3, drops from the "dun air sublime" near heaven onto what "seemed" the firm surface of the world, enclosed "[u]ncertain which, in ocean or in air" (3.72–76). Later, Satan is again sighted walking on the "firm opacous globe" (3.418). But the outer verge of the world appears much less "firm" after the text renames it a "windy sea of land" and likens the act of negotiating this zone to the flight of a vulture landing in a region best traversed by sailed wagons (3.431–40). As Satan passes from one cosmic place to another, the seeming boundaries bow and buckle, serving to divide light from darkness more effectively than distinguish kinds of matter.[54] Like advocates of the fluid-heaven theory and Torricelli in his letter to Ricci, Milton combines oceanic and atmospheric imagery to capture the materiality and fluidity of distant reaches of the sky and render a more physically unified picture of the universe than ever imagined before.

Yet there is no definitive evidence that Milton was aware of Torricelli's concept of the "ocean of elementary air." It is unlikely that he knew these specific words, or the original Italian, given that Torricelli did not personally publicize his experiment and only parts of his letter were passed on to Mersenne in France.[55] But there is one suggestive circumstance. The letter to Ricci quoted above was first published by Carlo Dati (1619–1676), a student of Torricelli's and Milton's closest friend in Florence. It appeared in 1663 in Dati's *Lettera a Filaleti*, a short work that promoted the scientific achievements of Torricelli including his experiment with mercury (see fig. 5). A well-regarded classicist and consummate scholar, Dati's interests were wide-ranging—he was appointed Chair of Humane Letters at the Florentine Academy in 1647–48 and was involved in the famous scientific society known as the Accademia del Cimento.[56] Dati, who was more than a decade younger than Milton and ever his admirer, penned a Latin tribute to the Englishman, and sent him an analysis of poetic diction as well as a copy of an oration he had given.[57] Milton reciprocated by sending him the Latin pieces from his recently published 1645 volume of *Poems*.[58] He was the only person from Italy with whom Milton corresponded after he returned to England, and since the two friends had traded work in the past, Milton seemingly would have been interested in Dati's publication on Torricelli, Galileo's intellectual heir, and his famous experiment.

Dati's book aside, it is virtually certain that in the years leading up to the publication of *Paradise Lost* Milton became aware of the Torricellian experiment. Bits of Torricelli's letter to Ricci and other accounts of the experiment,

LETTERA A FILALETI
DI TIMAVRO ANTIATE cioè del sigr Carlo Dati
Della Vera Storia della Cicloide, e della Famosissima Esperienza dell' Argento Viuo.

CARISSIMI FILALETI

Scriuo il vero à chi l'ama, e perciò senz'ornamenti, e senza lisci, sendo la verità tanto più bella, quanto più schietta, e più nuda. Socrate, che amaua questa nobil Donzella rifiutò la Difesa di Lisia, non come bugiarda, mà come troppo ornata. E Voi, o Filaleti, vi sdegnereste di sentir difendere il vero con artifici simiglianrissimi alla bugia. Difendo il Vero, mà senza maledicenze, perche la Verità si contenta d'essere impenetrabile, e rinunzia alla Menzogna le saette auuelenate del rancore, e della malignità. Imita ella generosamente la grauità imperturbabile degli Efori, quando fù loro, per racconto d'Eliano, e di Plutarco, bruttata di lordure la magistral residenza dalla insolente sfacciatezza de'Clazomeni, o de' Chij, ch'e' si fossero. Non s'adirarono essi, ma fecero per bando pubblico promulgar questo editto. Sia lecito a' Clazomeni operar bruttamente. O che bella vendetta! Così parmi adesso d'ascoltare la Verità oltraggiata sì, ma non irritata, che tranquilla, e ridente esclami ad alta voce. Tratti meco, e co'miei seguaci inciuilmente, e bugiardamente chi vuole, sopra di lui tornerà la vergogna, e l'offesa, come ricadono sopra la testa degli empi le saette, che s'auuentano contro al Cielo.

Dico adunque che agli anni à dietro uscì alla luce vn libretto scritto in Franzese, e intitolato, *Histoire De la Roulette*, e poi tradotto in latino, *Historia Trochoidis, siue Cycloidis, Gallicè la Roulette*; nel quale, a dire il vero, con maniere poco ciuili, e mal fondati argomenti, sendo intaccata l'ingenuità, la dottrina, e la riputazione, d'Euangelista Torricelli, Matematico, e Filosofo insigne del nostro secolo, e mio caro amico, e maestro, non potetti senza amarezza tollerare lo sfrontato ardire dello Storico, e poco mancò, che di subito io non prendessi la penna per redarguire fallacie così patenti. Ma poscia considerando, che tanti amici, e scolari del Torricelli, eguali d'affetto, e superiori di scienza poteuano ciò fare meglio di mè, mi ritirai per allora dall'impresa, alla quale ritorno adesso dubitando, che il silenzio rechi pregiudicio alla Verità, la quale è così chiara, e ben fondata, che non hà bisogno d'altra difesa, se non che chi la sà la disueli, acciò non resti adombrata dalle menzogne. Perciò fare non porterò sofismi, e chimere, ma testimonianze fedeli, scritture pubbliche per le stampe, e priuate originali, e autentiche, le quali saranno sempre esposte alla curiosità di chi volesse vederle; lasciandole considerare al retto, e spassionato giudicio degl'intelligenti, e de buoni, perche ne dieno diffinitiua sentenza. E mi protesto, che quando lo Storico, o altri replicassero a questa mia scrittura, per sostenere ostinatamente le loro proposizioni, io non farò giammai altra risposta, che questa, se però non mi capitasse qualche scrittura, o notizia di nuouo a fauor della Verità. Comincia per tanto l'Autore della Storia Cicloidale.

Inter infinitas linearum curuarum species, si unam circularem excipias nulla est, quae nobis frequentius occurrat, quam Trochoides, Gallicè, la Roulette. Vt mirum sit, quod illa priscorum seculorum Geometras latuerit, apud quos de tali linea nihil prorsus reperiri certum est. Describi-

A   *tur*

Figure 5. [Carlo Roberto Dati], *Lettera a Filaleti di Timauro Antiate; della vera storia della cicloide, e della famosissima esperienza dell'argento vivo* (Letter to the Filaleti from Timauro Antiate; regarding the true story of the cycloid and the well-known quicksilver experience) (Florence, 1663), 1. Timauro Antiate is Carlo Dati's pseudonym. (RB 701484, The Huntington Library, San Marino, California; photo courtesy of The Huntington Library)

ma persistesse nell'assenerare, che anche la natura concorren repugnare al Vacuo. Noi viuiamo sommersi nel fondo d' vn pelago d' aria elementare, la quale per esperienze indubitate si sà che pesa, e tanto, che questa grossissima vicino alla superficie terrena pesa circa vna 400. parte del peso dell'acqua. Gli Autori poi de' Crepuscoli anno osseruato che l' aria vaporosa, e visibile si alza sopra di noi intorno a 50. ouero 54. miglia; ma io non credo tanto, perche mostrerei, che il Vacuo douerebbe far molto maggior resistenza, che non fà, se bene vi è per loro il ripiego, che quel peso scritto dal Galileo, s' intenda dell'aria bassissima doue praticano gli vomini, e gli animali, ma che sopra le cime degli alti monti l'aria cominci ad esser purissima, e di molto minor peso, che la quattrocetesima parte del peso dell'acqua. Noi abbiamo fatti molti vasi di vetro come i seguenti se- gnati A, e B grossi, e di collo lungo due braccia; questi pie- ni d' argento viuo, poi serrata loro con vn dito la bocca, e ri- uoltatasi in vn vaso doue era l' argento viuo C, si vedeuano votarsi, e non succedere niente nel vaso che si votaua, il col- lo però AD restaua sempre pieno all' altezza d' vn braccio e 1. q. e vn dito di più. Per mostrar poi che il vaso fosse per- fettamente voto, si riempieua la catinella sottoposta d' acqua fino in D, & alzando il vaso a poco, a poco, si vedeua quando la bocca del Vaso arriuaua all' acqua descender quell'argento viuo del collo, e riempirsi con impeto orribile d' acqua fino al segno. E affatto. Il discorso si facena. Mentre il vaso A E staua voto, e l'argento viuo si sosteneua benche grauissimo nel collo A C, questa forza, che regge quell' argento viuo contro la sua naturalezza di ricader giù si è creduto fino adesso che sia stata interna nel vaso A E, o di Vacuo, o di quella roba sommamente rarefatta; ma io pre- tendo, che la sia esterna, e che la forza venga di fuori. Su la superficie del liquore, che è nella catinella grauita l' al- tezza di 50. miglia d' aria; però qual marauiglia è, se nel vetro C E, doue l' argento viuo non hà inclinazione, ne anco repugnanza per non esserui nulla, entri, e vi s' innalzi fin tanto, che si equilibri cõ la grauità dell'aria esterna, che lo spi- gne? l'acqua poi in vn vaso simile, ma molto più lungo salirà quasi fino a 18. braccia, cioè tãto più dell'argẽto viuo quãto l' argẽto viuo è più graue dell'acqua per equilibrarsi con la me- desima cagione, che spigne l'uno, e l'altro. Cõfermaua il discor- so l' esperienza fatta nel medesimo tempo col vaso A, e con la canna B, ne' quali l' argento viuo si fermaua sempre nel medesimo Orizonte A B segno quasi certo che la virtù non era dentro; perche più forza auerebbe auto il vaso A E, do- ne era più roba rarefatta, & attraente, e molto più gagliar- da per la rarefazione maggiore, che quella del pochissimo spazio B. Ho poi cercato di saluar con questo principio tut- te le sorte di repugnanze, che si sentono nelli varij effetti at- tribuiti al Vacuo, ne vi hò fin' hora incontrato cosa che non cammini bene; sò che à V. S. sonuer- ranno molte obbiezzioni, ma spero anche, che pensando le sopirà. La mia intenzione principale poi non è potuta riuscire, cioè di conoscere quando l'aria fosse più grossa, e graue, e quando più sot- tile, e leggiera con lo strumento E C, perche il liuello A B si muta per vn' altra causa, che io non credeua mai, cioè per il caldo, e freddo, e molto sensibilmente, appunto come se il vaso A E fussi pieno d'aria. Et vmilmente la riuerisco. Di Firenze 11. Giugno 1644.

Rispose il Ricci di Roma immediatamente sotto di 18. di Giugno 1644.

XXII. Il modo con che V. S. salua le esperienze fatte in riproua del vacuo, cioè del salire le cose graui contro sua naturale inclinazione, io lo giudico tanto più buono dell'altro, quanto che con questa ci conformiamo alla simplicità della natura nelle opere sue; la quale potendo saluare

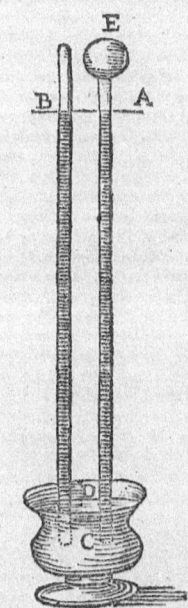

Figure 6. Page 21 of Dati's *Lettera a Filaleti*, which transcribes Torricelli's June 11, 1644, letter to Michelango Ricci, and includes a diagram of the so-called Torricel- lian experiment. Glass tubes of different sizes show that the vacuum within the vessels does not affect the height of the columns of mercury. (RB 701484, The Huntington Library, San Marino, California; photo courtesy of The Huntington Library)

which Mersenne himself had witnessed when he traveled to Florence in 1645,[59] inspired a series of demonstrations in France modeled on the Torricellian tube. Blaise Pascal (1623–1662), Florin Perrier, and Giles Persone de Roberval (1602–1675), among others, modified the experiment to prove that air may be condensed and dilated and that the pressure of the air decreases as altitude increases.[60] News of these French experiments met with excitement in England, notably in circles close to Milton. In 1648 Samuel Hartlib, the zealous promoter of civic projects to whom Milton addressed his 1644 tract *Of Education*, wrote to Robert Boyle with extracts of a letter from France that described "an experiment how to shew, as they suppose, that there is or may be vacuum."[61] Around this time, the scientific club known as the "1645 Group," said to have been initiated by the London intelligencer Theodore Haak, attempted the Torricellian experiment.[62] Haak, whose large acquaintance included both Hartlib and Milton, learned about the demonstration through Mersenne, whom he warmly thanked for the intelligence: "We remain, Sir, very much indebted to you for the communication of your experiment with the tubes and the mercury, for we have made two or three trials of it, in the company of men of letters and rank, with much pleasure and astonishment."[63] In 1648, the same year that Haak reported this activity to Mersenne, he informed Hartlib that Milton was writing a history of Britain.[64] If Milton discussed his writing project with Haak in 1648, Haak might well have mentioned something to Milton about his own ongoing engagements. He certainly had impressive news to share that very summer when his club demonstrated the Torricellian instrument to Lord Herbert of Cherbury and Prince Elector Palatine.[65]

In John Wallis's recollection, questions raised by the Italian apparatus loomed large in the London meetings of the 1645 group. So even if Milton was not the wiser for conversing with Haak, he might have heard, directly or indirectly, from the other enthusiasts in town who were occupied, in Wallis's words, with "the Weight of Air, the Possibility or Impossibility of Vacuities, and Nature's Abhorrence thereof" and "the Torricellian Experiment in Quicksilver."[66] Over the course of the next decade and a half, the mercury tube grew gradually more famous. Two important books printed in England in 1653 and 1654 publicized the debates associated with the experiment and gave detailed diagrams of the apparatuses concerned.[67] The central role of the tube of mercury in Robert Boyle's pneumatic research, unveiled in the 1660 volume *New Experiments Physico-Mechanical, touching the Spring and Weight of the Air*, brought the Torricellian experiment into the public eye.[68] In 1663, when the newly formed Royal Society was deliberating how to entertain Charles II for a planned visit the following year, Christopher Wren wrote to another member explaining why, lamentably, they lacked a suitable new experiment to show the king: "It is not every year will produce such a master experiment as the Torricellian, and so fruitful as that is of new experiments."[69] He emphasized that "the society hath deservedly spent much time upon that and its offspring," to justify why a substantially different experiment of the same caliber had not been delivered.[70] Having been in the limelight for almost twenty years and dominated the early research of the Royal Society, the Torricellian experiment was too well known in 1664 to be trotted out again. Living in London throughout this period among people such as Lady Ranelegh, who knew the Oxford experimentalists intimately, Milton must have taken an interest in the "master experiment" of the Torricellian tube. Indeed, it would have been impossible to ignore this instrument of the Galilean school, especially because it evinced a wholly novel definition of air, the element in which human life thrives and is, apparently, submerged.

PARADISE LOST AND THE AIR'S "UNUSUAL WEIGHT"

Capability to stay suspended in the air versus falling down is a moral index in *Paradise Lost* of material things, beings, and the narrative itself. Weight plays a crucial role in determining this capability. As John Rumrich argues, the association of the Hebrew word *kabod* in scripture with weight or substantiality shapes the concept of glory, a defining attribute of Milton's God and theodicy. Through their finite substantiality created beings in *Paradise Lost* partake in the infinite material essence of God.[71] My interpretation of the epic affirms the relationship of weightiness to the actualization of God's substance in the world. I make a further point that explains the fluctuating perception of weight in the epic and seeming reversals between the moral values of heavy and light. In addition to Hebraic and Aristotelian ideas of weight, the epic draws on the concept of air pressure laid bare by the Torricellian experiment to portray the experience of feeling and wielding weight in a universe that is undergoing climate degradation and elemental recalibration. By rendering atmospheric pressure visible, but also obliterating it in the void space, pneumatic instruments confirm in some cases that an object's lightness or heaviness is relative—dependent on invisible forces in the atmosphere. *Paradise Lost* alludes to the scenario of the Torricellian experiment to liken Satan's devastating rise as "Prince of the air" to the manifestation of pressure in the atmosphere and its destabilizing effect on the concepts of heavy and light (10.185).

As readers have often noticed, the phenomenon of flight, whether human, spiritual, imagined or real, is a favorite topic in *Paradise Lost*.[72] But depicting flight as a means of conveyance does not always seem to be the point of such representations. Rather suspension, that is, hovering or hanging in the air, is often the overriding effect. In the opening lines of the epic, for example, the narrator states that he intends his "advent'rous song" to soar "with no middle flight" above the summit of pagan literature (1.13–14). Achieving a specific altitude, metaphorically speaking, is more important than going to a specific place. A few lines later, the narrator enlists the power of the Holy Spirit to lift up his baser parts: "what in me is dark / Illumine, what is low raise and support" (1.22–23). The epic song is thus fashioned as a product of two collaborating agencies: one is naturally weighty and earth-bound, but seeking to rise up, and the other is heavenly and capable of promoting what is earth-bound to an unnatural height. Divine frustration of natural law—at least, the natural motions Aristotle assigns to light and heavy bodies—opens the way to prophetic vision and song.

Satan's predicament at the beginning of book 1, ironically, is similar to Milton's. He can do nothing, let alone embark on his long and arduous flight, without first rising up. Lying flat and chained on hell's burning lake, he cannot deliver himself without "supernal" approval (1.241). As the narrator points out, Satan never would have

> risen or heaved his head, but that the will
> And high permission of all-ruling heaven
> Left him at large to his own dark designs.
>
> (1.211–13)

Just as Satan is on the verge of getting up, the epic further justifies why heaven allows it. God does not interfere with Satan so "[t]hat with reiterated crimes he might / Heap on himself damnation" and "[t]reble confusion, wrath and vengeance poured" (1.214–15, 220). The weight-bearing verbs "[h]eap" and "pour" dampen the iambic verticality of the line, "Forth*with* up*right* he *rears* from *off* the *pool*" (1.221; emphasis added). Satan propels himself upward off the lake only to load himself with additional guilt.

The verbal contrast between descriptions of Satan picking himself up and the imminent downpour of damnation reflects the interplay between two physical properties of air—elasticity and pressure—recently discovered by the Torricellian experiments. Milton's use of the terms *pour* and *heap* recalls Pecquet's characterization of the atmosphere, quoted above, as "a heap of spongious, or rather wooly material." In the translation of Pequet's book, which discusses certain French trials of the Torricellian experiment, the term "heap" is repeated a few lines later in the section that introduces the concept of the air's "elater"—what Boyle later calls the "spring of the air."[73] He writes: "And for that cause by its Spontaneous dilatation [which I call Elater] howsoever the *heap'd on burthen* do press them down, yet the under parts, if they have liberty, will endeavor to rarifie themselves."[74] The reference to the parts below the heaped on burden of superimposed atmosphere, which, if freed, express elatery or propulsion, distinctly parallels Milton's description of Satan, who "[l]eft . . . at large," quickly rears himself and exits the lake despite the treble burden that awaits him.

What we might call the "lift-off" theme of book 1 echoes the discourse about the Torricellian experiment in focusing on the idea of suspension and explaining its cause. Torricelli in his original letter to Ricci correctly explained how his device worked, but questions still dogged the experiment. A main sticking point for contemporaries was identifying the force that holds the column of mercury up. In the seventeenth experiment of the *New Experiments Physico-Mechanicall,* Boyle sets out to settle the question by placing a Torricellian tube inside of an air-pump.[75] The logic behind the experiment is this: if the suspension of the column of mercury in the Torricellian tube is caused by the atmosphere pressing on the mercury in what Boyle calls the "subjacent"[76] dish, then evacuating the receiver of any air should lower the column of mercury and empty it into the dish. A predecessor called this kind of set up "the Experiment of Vacuity in Vacuity."[77] Boyle thought of it "as if we were allow'd to try the Experiment beyond the Atmosphere."[78] If Boyle managed in 1660 to conduct the Torricellian experiment in conditions like those "beyond the Atmosphere," that we should find signs of it in Milton's hell seems all the more plausible.

Boyle and Hooke performed the experiment and confirmed that sucking the air out of the receiver by turns gradually lowers the column of mercury. Then they ran the experiment backwards: "[W]e Returned the Key and let in some new Air; upon which the Mercury immediately began to ascend (or rather to be impell'd upwards) in the Tube, and continu'd ascending, till having Return'd the Key it immediately rested at the height which it had then attain'd: And so, by Turning and Returning the Key, we did several times at pleasure impel it upwards, and check its ascent."[79] It would be theologically problematic to suggest that God pushes Satan up, or by analogy, opens the valve of the air pump and provides the air that rushes into the receiver. That God might use his breath to blow the fires of the burning lake "into sevenfold rage / and plunge [the devils] into the flames," is possible according to Belial; but this is a far cry from using his breath to actively restore Satan and send him on his mission of destruction (2.171–72). And yet there is a troubling symmetry between the image of Satan's "mighty stature" rearing "from off the pool," and the column of mercury in the Boyle's tube that "immediately began to ascend (or rather to be impell'd upwards)" off the pool of mercury in the subjacent dish (1.222).

The description of Satan in the beginning stages of flight include a further allusion to pneumatics:

> Then with expanded wings he steers his flight
> Aloft, incumbent on the dusky air
> That felt unusual weight, till on dry land
> He lights.
>
> (1.225–28)

David P. Harding discerns in these lines a parallel to Virgil's sea serpents, which "press down the sea" (*incumbunt pelago*) as they swim to shore where they kill Laocoön and his sons.[80] But the fact that Satan's body presses down on the air, not the lake, implies an update to Virgil's passage that reflects contemporary insights about the atmosphere and the new language of pneumatics.

The key word in the passage is "incumbent," an Anglicization from Latin that entered the English language, probably during the early seventeenth century. It describes that which "lies, leans, rests, or presses with its weight upon something else."[81] This much we can learn from an editor's gloss. But none of the standard editions (Verity, Elledge, Hughes, Fowler, Flannagan, Leonard; Kerrigan, Rumrich, and Fallon) mentions seventeenth-century physics as a potential context for the term. Yet, in the proliferating discourse of pneumatics, "incumbent" is the word preferred by Beeckman, Pecquet, Walter Charleton and others for describing the pressure exerted by "air, fluid, or other weight."[82] It communicates that air is a body that can lean or press with its weight and suggests the effect of the higher or "incumbent" part of the atmosphere on the compressed layers below it.

According to the *OED*, "incumbent" was first used to express the pressure of air in Boyle's *New Experiments Physico-Mechanicall*.[83] But in fact it appears in earlier publications that discuss the phenomenon of air pressure. Beeckman who pioneered the idea of air pressure and analogized the atmosphere to a sponge, was possibly the first person to use the Latin equivalent in this specific context, writing in the "Corollaria" to his MD thesis in 1618 that water sucked up by a pump is not pulled by a vacuum, but rather pushed up by the *aere incumbente*—the pressing or leaning of the air.[84] In his discussion of air pressure and the Torricellian experiment, Pecquet remarks that the parts of the air nearest the earth "are compressed compactly by the weight of the parts incumbent."[85] Published the following year, Walter Charleton's *Physiologica Epicuro-*

Gassendo-Charltoniana, which includes a twenty-five-page discussion of the Torricellian experiment, similarly describes the phenomenon of air pressure as the "[d]epression of the inferior parts of the aer by the superior incumbent upon them."[86] Charleton also refers to the portion of the atmosphere whose weight balances against the mercury in the tube as "the impendent Cylindre of Aer" (see fig. 7).[87] However, "incumbent" seems to be more common than "impendent." By the 1660s when Boyle presented his air-pump experiments, the word "incumbent" had been a fixture of the topic long enough that English synonyms of the Latin word, such as "leaning," "pressing," and "pressure," were also becoming staples of the discourse. Boyle uses this tree of related words to explain that air pressure inside the receiver is constant with that of the ambient air: "[W]hen the Air is shut into the Receiver, it does . . . continue there as strongly compress'd, as it did whil'st all *the incumbent Cylinder of the Atmosphere lean'd immediately upon it*; because the Glass, wherein it is pent up, hinders it to

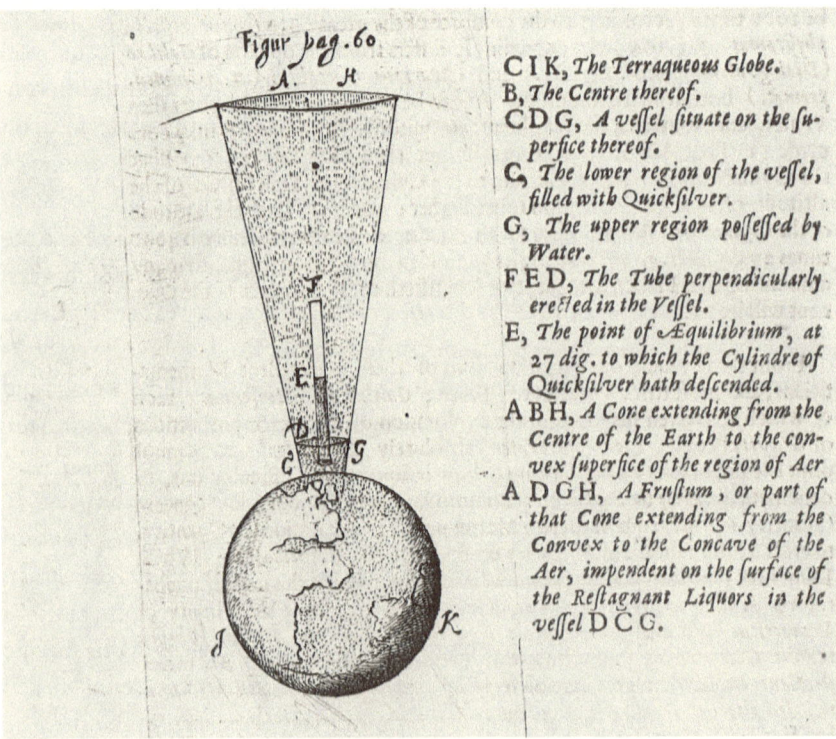

Figure 7. Barometer and incumbent or "impendent" atmosphere, in Walter Charleton, *Physiologia Epicuro-Gassendo-Charltoniana: or a fabrick of science natural, upon the hypothesis of atoms* [. . .] (London, 1654), 60. (Photo courtesy of the Wellcome Collection)

deliver itself, by an expansion of its parts, from *the pressure* wherewith it was shut up."[88] By the publication date of *Paradise Lost,* the term "incumbent" was not only notionally linked to the weight of the atmosphere and compaction at its base, but also used in tandem with "pressure," the modern scientific measure of such compaction.

It is in the pneumatic sense, then, that Satan is "Aloft, incumbent on the dusky air." Since he is "[a]loft," or high up in the air, he is analogous to the superior atmosphere that, as commentators on the Torricellian experiment explained, leans on and compresses the lower parts of the air. His body produces air pressure. As a result, Satan rests on air "[t]hat felt unusual weight, till on dry land / He lights." The exact import of the phrase "unusual weight" in these lines is slightly ambiguous. Does it signify that Satan's body is extraordinarily heavy, or that *any* burden incumbent on the air in hell is unusual? The apparent source of these lines adduced by Robert Thyer, the only critic to suggest a precedent that I am aware of, stresses the idea of uncommon heaviness instead of the unusualness of weight per se. In a note contributed to Thomas Newton's eighteenth-century variorum edition of *Paradise Lost,* Thyer suggests that Milton draws on Spenser's description of the old dragon in *The Faerie Queene,* book 1, canto 11, stanza 18.[89] The comparison is reasonable. Later in the epic, Satan is associated with the dragon of Revelation[90] and is massive like Spenser's monster, whose flight

> did forcibly divide
> The yielding aire, which nigh too feeble found
> Her flitting partes, and element unsound,
> To beare so great a weight.[91]

But the two passages differ in crucial respects. In contrast to Spenser's description, which focuses on the disintegration and failure of the "unsound" air under the violence of the dragon's wings, Milton portrays Satan's ascension as a physical and environmental novelty. The previously unburdened air becomes compressed under the weight of Satan's body. It is not in danger of giving way, but as Joseph Addison said of Milton's effect on the English language: "it sunk under him."[92] This may imply Satan's uncommon heaviness, but the key issue is the novelty of weight itself, which Milton describes as being "felt" by the air like a new sensation.

Milton underscores novelty in this episode, I propose, because the idea that air has any weight at all was still striking and new. Only recently, during the interval between Milton's young adulthood and middle age, had evidence of the incumbent "ocean of elemental air" come to light and dislodged Scholastic notions about the air. Milton's contemporaries, in a sense, "felt unusual weight" in the air. The technology and vivid imagery of the pneumatic experimentalists with their analogies to sponges, springs, and wool, rendered the physical heft and pressure of the air both visible and real. Beale's awestruck exclamation indicates the extraordinary impact of this innovation: "who could ever expect, that we men could find an Art, to weigh all the Air that hangs over our heads, in all the changes of it . . . ?"[93] Having witnessed this conceptual revolution flowing from Galileo's assertion that air has weight and Torricelli's demonstration of atmospheric pressure, Milton draws on its staggering conclusions to develop the idea of original climate change in the epic. Chapters 3 and 4 argue that as Satan gradually assumes his scripturally appointed role as "Prince of the air" (10.185) he promotes the physical and moral pollution of the air and hastens the climatological devastation that accompanies the Fall. In light of this broader association between atmospheric degradation and the corruption of mankind, the scene in book 1 appears all the more ominous. In addition to introducing inclement weather, noxious fumes, and deceptive sounds into the fallen atmosphere, Satan is presented in this originary act of flight as the author of the air's weight and pressure, a physical manifestation of the moral burden of sin.

THE "PRINCE OF THE AIR" ON THE SCALE

In the same way that Satan can make the weather, but is also "weather-beaten," his relationship to atmospheric weight is contradictory (2.1043). On the one hand, he embodies a heavy burden to be imposed on Earth's region of air, but at the same time is ruled by oscillations in his environs. At the council of hell, Beelzebub suggests that it is within Satan's power to make himself virtually light even though he bears a great weight. Speaking implicitly about Satan, Beelzebub imagines that whoever journeys to Earth will cross the abyss of chaos in "airy flight / Upborne with indefatigable wings"—in spite of carrying on his back "the weight of all and our last hope" (2.407–8, 416). In reality, however, brute force and willpower are *not* the factors that see Satan through the gulf. To be "[u]pborne" in "airy flight" requires the chance cooperation of the environment as when Satan plummets through chaos only to shoot upward moments later on a rising cloud (2.931–38).

Satan attempts a form of barometry to find a safe passage to Earth. This becomes evident if we understand that the Torricellian barometer is essentially a scale or balance. Indeed, the barometer was also known as the "Ballance of the Air," and Boyle called one of his variations on the Torricellian tube, the "scale-barometer."[94] The reason it was known as a scale is that the mercury in the tube descends to the height at which its weight perfectly balances against the pressure produced by weight of the incumbent atmosphere. Walter Charleton refers to this point as the "*Æquilibrium*" or "*Æquipondium*," emphasizing the equality of weight between the air and the suspended mercury.[95] Wishing to hover in equilibrium with his surroundings, Satan also compares his weight with atmospheric conditions. Whether like a fleet seen far off he "*[h]angs* in the clouds," or stands on the border of chaos, "*[p]ondering* his voyage," or reaching "the emptier waste, resembling air, / *Weighs* his spread wings," Satan continually takes pressure readings—checking that his weight will be supported by the elements (2.637, 919, 1045–46; emphasis added).

A well-known episode in book 4 further illustrates the idea that Satan's weight may be reduced, even completely obviated, by divinely augmented spiritual and environmental forces. When the captured fiend challenges Gabriel whose strength God has "doubled" and God's golden scales decide the outcome, the seemingly heavy Satan is "weighed, and shown how light" by the uplifted arm of the scale (4.1009, 1012). It is crucial to note, as the narrator points out, that these are the same golden scales

> Wherein all things created first he weighed,
> The pendulous round earth with balanced air
> In counterpoise.
>
> (4.999–1001)

This insertion identifies the scales that decide Satan's fate with the scales of Creation, establishing an ominous echo between the original perfection of the element of air and the weighing of Satan, future prince of the air.

Previous commentary on the scales has focused on their literary and astrological import. Their meaning in light of natural philosophy, a domain where the balance played a major role, has been less adequately explained. In a rare gesture to the scientific context of this episode, Harinder Marjara argues that this description of global equilibrium alludes to philosophical traditions that posited the earth's steadiness on its own axle or as a point within the universe. But Marjara concludes that the image of the air and earth balancing on a scale "does not owe to any specific source."[96]

But in fact we do not need to look far for a precedent. Milton's contemporaries imagined putting earth and air onto a balance and, in at least one case, they tried doing it. The dream of lifting the whole weight of the world originated with Archimedes, who once joked that he could do it "given a place to stand."[97] The Archimedean proposition, which implied the use of a lever, served during the seventeenth century as an exemplum of the miraculous potential of mechanics and mathematics. Wilkins's *Mathematicall Magick*, for example, includes a diagram of the globe poised on a lever, about to be lifted. Above this arrangement, another "world" hangs pendulum-like from the short arm of a balance (see fig. 8).[98] Weighing the earth as a single entity was still an utterly fictional idea, but such depictions did not approach it as a philosophical impossibility. The optimism of the New Science infuses Milton's image of the scales even though it includes the important caveat that the Deity, not a human being, lifts the "pendulous round earth."

The history of pneumatic investigation further demonstrates the scientific coherence of the whole image and explains the presence of air on the scale. The idea of causing the air to equiponderate with earth was utterly germane to the Torricellian experiment and was attempted in the most literal manner by Boyle. In his thirty-fourth experiment with the air pump, a balance is used to

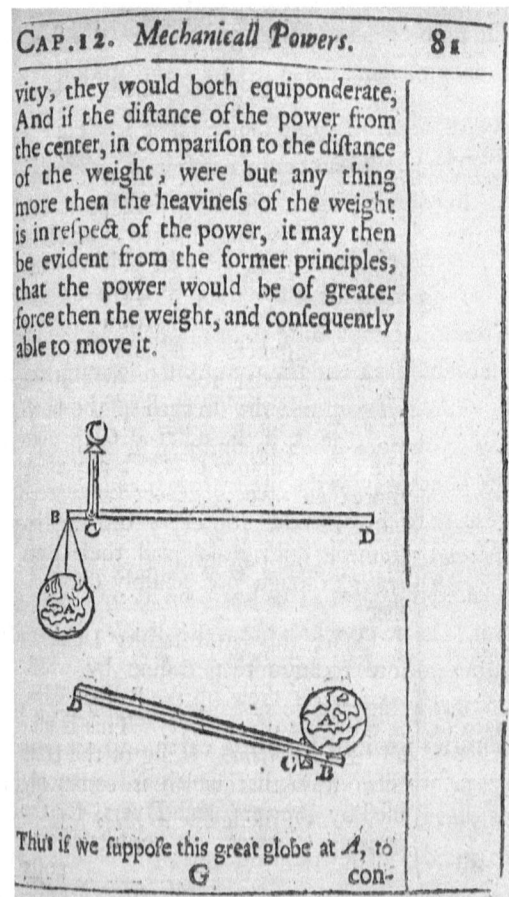

Figure 8. Woodcut showing how a balance and lever might be "contrived to move the whole world," in John Wilkins, *Mathematicall Magick. Or, the Wonders that May Be Performed by Mechanicall Geometry.* [. . .] (London, 1648), 1:81. (RB 487000:0437, Mount Wilson Observatory Collection, The Huntington Library, San Marino, California; photo courtesy of The Huntington Library)

weigh the air against a metal counterweight. Boyle's description of the experiment anticipates Milton's diction of weighing, balancing, and counterpoise. According to Boyle's account, they fastened a bladder "about half fill'd with Air" to one scale, and "put a Metalline counterpoise into the opposite Scale."[99] When the two objects were "brought to an *Æquilibrium,*" or shown to weigh the same, "the ballance was convey'd into the Receiver" of the air pump.[100] As they reduced the air pressure in the chamber, the bladder began to inflate and appeared "to outweigh the opposite weight, drawing down the Scale to which it was fastned very much beneath the other, especially when the Air had swell'd it to its full extent."[101]

The bladder's expansion under reduced air pressure and its "drawing down" of the scale suggest the springlike action of the air, which Boyle also refers to as its "Power of Self-Dilatation," or the tendency of the corpuscles of air to stretch or expand outwards.[102] The phrase "Self-Dilatation" closely aligns with Milton's description of Satan's fearsome posture as he prepares to fight. With the angelic guard standing across from him, as if already balanced on the scale,

> On the other side Satan alarmed
> Collecting all his might *dilated* stood,
> . . . .
> His stature reached the sky.
> (4.985–88; emphasis added)

Instead of pushing down the balance like the dilated bladder, however, Satan sees his "mounted scale aloft" (4.1014). His fate matches "the flagging of the bladder," the deflation that follows self-dilatation as air is let back into the evacuated receiver and normal pressure is restored.[103] Likewise, in the passage from *Paradise Lost,* a divine influx of wind "bends" and "[s]ways" the angelic guard like ears of corn, guiding their conduct towards the devil (4.981, 983). The bending of the "bearded grove," resembles the constriction of springs or fibers of wool, signaling the compression of air under the incumbent atmosphere (4.982). If the battle between Satan and Gabriel is averted by divine permission, the action that allowed Satan to exit the burning lake in book 1, then a pneumatic reading of the passage suggests that God assents to Satan's going free, ironically, by turning up the pressure. God assumes the compressive function of the incumbent atmosphere as he pleases, changing the dynamics of the atmosphere to diminish Satan's dilation or apparent heaviness, but also release him to pursue further damnation.

Before building pressure produces a new equilibrium between the elements at the Fall, gentler forces hold together the universe. As we have seen, during the creation of the world the air and the "pendulous" earth were brought into perfect balance. The image of the scales weighing air and earth portrays the atmosphere in suspension, hovering as on a scale rather than bearing down on the earth. The atmosphere's amity with Earth's creatures promotes its lightness so that "the air / Floats" as a flock of birds migrates through it, "fanned with unnumbered plumes" (7.431–32). The elemental pair of air and earth, moreover, are depicted in spousal terms. The earth softly carries "the smooth air along" as it turns, and the "ambient air" is likened to enfolding arms, "[e]mbracing round this florid earth" (8.166, 7.89–90).

But this endearing relationship ceases after God's angels bring a curse against "heaven and earth" causing the warring winds to scar the face of the land with opposing blasts (10.638–40, 695–706). As Adam watches the sky and wilderness change and wonders if he is witnessing the "end / Of this new glorious world," he feels corresponding and even worse "miseries" gathering within him (10.715, 720). It dawns on him that he will be reviled by his offspring for causing their suffering (10.733–35). Suddenly the curse multiplies into curses, all falling back on him:

> so besides
> Mine own that bide upon me, all from me
> Shall with a fierce reflux on me redound,
> On me as on their natural centre light
> Heavy, though in their place.
>
> (10.737–41)

The comparison plays on the axiom from Scholastic physics that a body loses the inclination to move and thus becomes weightless when it gravitates to its natural place. According to the old physics, the material curses of Adam's offspring, which recoil back on their common ancestor or "centre" should not have any weight. Assuming perhaps that Milton accepted the Scholastic rule, Marjara views the simile as a device for differentiating the heavy burden of the curses from the presumed "lightness of the elements in their domain."[104] But confronted with the counterevidence of the Torricellian experiment, Milton and his contemporaries knew that the elements weigh in their natural places, even the element of air. Adam too has just witnessed the deterioration of the climate and appreciates that a grave change has occurred. The Argument's summary of the moment when sin sinks in, "Adam *more and more perceiving his fall'n condition heavily bewailes*," describes the act of lamenting as heavy itself ("bewail" is used in the intransitive form). Thus Adam's complaint that his progeny's hatred will "redound" on him, though it conveys nostalgia for a world wherein the elements are "light" in their spheres, acknowledges in a very literal sense the gravity of the new fallen dispensation: curses made of air truly are heavy burdens.

We have seen in previous chapters that *Paradise Lost* portrays the first humans in the original instance of sin acquiescing to Satanically corrupted sounds and meteorological phenomena, and thus bringing upon the race a permanent climatological curse. To the list of atmospheric defects Milton associates with this judgment may be added air pressure and weight, properties of the fallen climate proven to exist during Milton's lifetime that embody the physical occupation of the atmosphere by Satan and his legions after the Fall. As we shall see in the next chapter, in spite of mounting scientific evidence of the mechanical behavior of air, *Paradise Regained*, the sequel to *Paradise Lost*, represents corrupted air as the legacy of sin and a root problem for human history and theodicy. Rather than ignoring or refuting new scientific insight on the air, however, *Paradise Regained* incorporates such knowledge to suggest that environmental conditions like those of Restoration London perpetuate the original degradation of the climate. The temptations of Jesus illustrate the role of great human civilizations—in cooperation with Satan—in further aggrieving the corruption of the air. Rehearsing a possible future resolution to the scourge of atmospheric corruption, the words and actions of the hero of Milton's shorter epic epitomize humanity's potential to stave off and counteract the climate's heaviest assaults.

CHAPTER 6

"Throttled at Length in the Air"
Environmental Warfare and Climate Regained

When the volume containing the poems *Samson Agonistes* and *Paradise Regained* was entered into the Stationer's Register in September of 1670, Milton's England had already for a decade reverted to monarchial rule.[1] But it was still a fractured nation fraught by, among other tensions, the prospect of a Catholic successor to the throne and the ongoing persecution of nonconformists.[2] In the shadow of the Licensing Act of 1662, much of the political and religious debate that had been aired openly in the press during the Interregnum moved into unregulated spaces such as the coffee house.[3] After the new regime briefly imprisoned Milton, his Republican pamphleteering ground to a halt for the duration of the 1660s.[4] The pair of poems, published in spite of this climate of censorship and hostility toward the Good Old Cause, depicts the biblical figures of Samson and Christ combatting spiritual and cultural opposition. Seen as political allegories with unmistakable parallels to Milton's difficult post-Restoration years (as these poems are often read) they explore the chasm between the attitudes of a corrupt ruling party and the ideals of a righteous dissident.

Yet, we cannot let Milton's stinging political defeat deafen us to the ecological distress that also rings out in the poetry of this era. Encompassing and echoing the poet's political anguish of the 1660s and '70s, the problem of the oppressive physical climate and *its* effect on the moral and material welfare of humanity looms large in these poems. Yet critics have not directly addressed this crucial theme. In *Paradise Regained*, in particular, climate defines the central conflict. It constitutes the grounds and prize of the spiritual battle as well as the primary source of Satan's power. This chapter reframes the conflict of the brief epic, which has typically been treated as clash of two contending intellects, in environmental terms. Jesus confronts the demonic control of the fallen climate and the moral quandary of subsisting on its fruits. The dissonance between the hero's internal calling and external temptations, a contrast explored in *Samson Agonistes* as well as in *Paradise Regained*, is a symptom of the global problem with which Milton's poetry is constantly wrestling, that of being human and occupying a climate severely corrupted by the Fall.

Interpreting the contest in *Paradise Regained* as a struggle between the human subject and the compromised environment clarifies the implications of original climate change for Milton's concept of free will. This study has thus far suggested there is scant breathing room in the fallen climate for exercising one's will freely. Obtruding sounds and airs corrupted by malicious spiritual agents in *Paradise Lost*, for example, constrain Eve's deliberations about whether to eat the forbidden fruit. Milton's core belief in the unitary material basis of body and soul lies behind this portrayal of the will as vulnerable to environmental influences. John Rumrich and Stephen M. Fallon argue that Milton's monism refuses to wall off human consciousness from the environment, claiming that a major effect of the epic poetry is "to expose such a division between self and the world as a fantasy."[5] The porosity of the self-world boundary can facilitate harmony throughout God's creation as in prelapsarian Eden. But the stakes change when the atmosphere is corrupted. Scholars have not adequately recognized a major consequence of Milton's particular strain of materialism: the embodied soul and its spiritual foes are physically enmeshed in the world and, thus, potentially communicable with each other. In Milton's monist universe, matter feeds the human body's intellectual spirits, which, in turn, comprise the soul.[6] Thus even the body's physiology potentially exposes the soul to malignant influences implicit in the environment. To complicate matters further, the animistic environment itself is willed and thus susceptible to sin, introducing additional opportunities for dissembling and indirection between the satanic tempter and the human soul.[7] With the effects of original sin latent in the very climate, volition for Milton consists not only of good discernment, but also bodily hardiness and resilience. Nowhere in Milton's poetry is this more apparent than in *Paradise Regained* where physical attributes such as self-denial, vigor, and balance, are so baldly put to the test.

In this brief epic, Christ occupies a monist materialist universe, and his heroism is thus shaped by its postlapsarian ecology. To overcome the enemy without becoming compromised himself, Christ is called to perform something apparently miraculous, though narrowly within our reach. He obstinately deflects and counteracts, time and time again, climatic influences that are almost impossible to disentangle from the human organism, embodying an ethic of forbearance, which, if sustained, can expel atmospheric corruption from the fallen world.

SATANIC METEOROLOGY AND LONDON'S AIR POLLUTION

The deterioration of London's air, which began decades before the publication of the 1671 poems, and the perceived forces behind this climate change, shed light on Christ's combat with atmospheric corruption in *Paradise Regained*. Though it is sometimes supposed that London's characteristic smog arrived with the machines and manpower of industrialization, plumes of smoke from ordinary chimneys were already widespread by the beginning of the seventeenth century, and air conditions became abysmal toward the end of Milton's lifetime as the city accelerated its consumption of coal.[8] Emily Cockayne has shown that the simple act of breathing or smelling the air in mid-seventeenth-century London was a disgusting, if not unhealthful, experience.[9] Contemporaries balked at the poor air quality and debated the reason for the suffocating conditions. John Evelyn incorrectly blamed London's brewers rather than the hearth fires of its private citizens for dirtying the air in his 1661 tract *Fumifugium,* but he correctly articulated the essence of the crisis: the burning of coal submerged the city in an intolerable cloud of smoke that endangered public health, and, as he argued, deserved the attention of the highest levels of the government.[10]

An increasingly offensive feature of the urban landscape during Milton's life, the problem of coal pollution in London traces back approximately to the time of his birth. During the late sixteenth and early seventeenth centuries, England switched from burning wood to coal—specifically, "sea coal," a cheap fuel with a high sulfur content, which was shipped south from the Newcastle region to London where it dominated the market.[11] In the middle of the sixteenth century, 10,000–15,000 tons of coal were imported annually to London, whereas by the late 1680s, in certain years the Thames saw a half a million tons of coal cross its banks.[12] Drawing on a scientific model that estimates average annual levels of pollutants in London's atmosphere from the twelfth century forward, a recent study of London's smoke crisis shows that "[t]he air of England's capital during the early modern period . . . was roughly as polluted as the very dirtiest cities in our contemporary world."[13]

Milton lived in or around London's Barbican neighborhood periodically throughout his adult life: from 1645 to 1647, during the first half of the 1660s, and again (after a period away from the city) from 1669 until his death.[14] In this populous and largely poor northeastern part of town, he would have had to cope with relatively severe effects of the air pollution problem, which was concentrated away from the wealthier West End and nearer the crowded city center with industries along the south bank.[15] In 1666, smoke became a nuisance for Milton's one time neighbor and patron the Earl of Bridgewater who owned a large house in the Barbican very close to where Milton lived in the mid-1640s. A neighboring soap boiler's new chimney began to blow pollution into his property, and Bridgewater took steps to put a stop to it.[16] Harm to health or private property caused by someone else's smoke was among the many kinds of nuisances eligible for redress under early modern common law, where "nuisance" (though it did not have a stable legal definition) generally meant annoyance, inconvenience, or damage.[17] Affecting the sense of smell and respiration, polluted air was probably a particular irritant for the blind Milton whose mode of interfacing with the world in his later years depended entirely on nonvisual sensation. As Samson complains, blindness makes a person "obnoxious more"—further exposed—to life's "evils, pains and wrongs" (*Samson Agonistes*, 105–6).

Prior to 1660, Milton had written poetry that depicted the postlapsarian climate as an instrument of moral punishment. In the Restoration's opening decade this idea harmonized almost prophetically with both the quality of the air in London and the disastrous events that played out against this smoky backdrop. In addition to enduring a demoralizing political defeat, Milton and his family were forced to flee the plague in July of 1665 and thereby avoided the catastrophic Great Fire of 1666, which consumed much of the ancient walled city of London.[18] We can safely assume that this poet, who was already wary of the effects of original sin in Nature, drew a line between the city's offensively smoky (and apparently pestilential) air and the death and misery that reigned during this brief interval of history. Scholars have begun to assess Milton's engagement with this ecological crisis in *Paradise Lost*. Ken Hiltner argues that references to sulfur and smoke in Milton's portrayal of hell build on a longstanding literary trope that fused key traits of England's sea coal pollution, such as its odor of sulfur, with descriptions of hell. According to Hiltner, the association grew until the English conception of hell acquired a sulfurous foundation—"brimstone"—and "[air pollution from sea coal] now seemed hellish."[19] Hiltner's astute comparison of the mining activities depicted in hell to London's coal supply evinces Milton's awareness of the environmental costs of the urban fuel crisis. This instructive interpretation, however, misses the crucial parallels that the poetry draws between the mining of infernal coal, Satan's office as prince of the air, and various kinds of meteorological corruption that demonic agents perpetuate in the fallen world.

In addition to alluding to London's air pollution problem, Milton's depiction of the mining of sulfur in hell prefigures Satan's role as a weather maker in the fallen world. The seventeenth-century's "gunpowder" or "sulfur-nitre theory" of meteorology, so named by the historian of science Henry Guerlac, proposes that the interaction of two main ingredients of gunpowder, sulfur and nitre, potentially causes thunderstorms and earthquakes as well as other phenomena such as snow and hail.[20] The reaction of these substances might occur above or beneath the ground. The sulfur-nitre theory influenced several major Continental and English philosophers, including Daniel Sennert, Thomas Browne, and Isaac Newton.[21] Comparing the report of firearms to the sound of thunder, for example, Browne attributes "those terrible cracks, and affrighting noyses of Heaven" to "nitrous and sulphureous exhalations, set on fire in the clouds" that break free and lacerate the air.[22] Building on the analogy, he goes on to speculate that "subterraneous Thunders and Earthquakes" occur "when sulphureous and nitrous veins being fired upon rarefaction, do force their way through bodies that resist them. Where if the kindled matter be plentiful, and the Mine [i.e., mineral] close and firm about it, subversion of Hils and Towns doth sometimes follow: If scanty, weak, and the Earth hollow or porous; there only ensueth some faint concussion or tremulous and quaking Motion. Surely, a main Reason why the Ancients were so imperfect in the Doctrine of Meteors, was their ignorance of Gun-powder and fire-works. Which best discover the causes of many thereof."[23] For Browne, the transformation of minerals into the volatile substance of gunpowder illuminates a common cause of several kinds of meteors. Ignited by sulfur and nitre circulating above and below the ground, meteorological phenomena differ mainly in their location; "terrible cracks, and affrighting noyses of Heaven" are the atmospheric counterparts of earthquakes or "subterraneous Thunder." Surpassing Browne's "Ancients" in knowledge both of gunpowder and meteors, Satan realizes that the key to artificially reproducing heavenly thunder lies in the "sulphurous and nitrous foam" buried underground (6.486–91, 512). In the infernal pit, he finds plenty of these weather-making materials. Compared in a famous simile to "thundering" Mount Etna, the very soil of hell teems with them (1.232–38). As Guerlac speculates, the "combustible / And fueled entrails" in the analogy allude to nitre and sulfur, which were thought to cause earthquakes and volcanic eruptions (1.233–34).[24] Nor is the chemical lake of hell, fed with "ever-burning sulphur unconsumed," lacking meteorological fuel (1.69).

In the concluding scene of book 1, the devils' excavation of a volcanic hill enclosing gold—"[t]he work of sulphur"—as Hiltner argues, resembles a coal mining operation, and the smoke above the hill's "grisly top" the associated emissions (1.670, 674). However, the epic's pointed allusions to the gunpowder theory of meteors, the abundance of sulfur in hell, and Satan's other attempt to excavate the ingredients for manufacturing thunder—suggest that the ore extracted from this sulfurous hill also represents the germs of storms and natural calamities. The fallen angels' environmental intervention thus portends the spread of sea-coal pollution in seventeenth-century London, and by the same token, satanic corruption of meteorological phenomena *in general*. More than merely polluting the air, the act of mining coal or other substances, enables a permanent climate change—a satanic turn—in the character of the weather.

If Milton's characterization of the devils was influenced by London's air pollution problem, it was not solely because of the suggestive subterranean derivation of coal. The doctrine of the "prince of the power of the air," which grants that Satan uses meteorological effects to frighten and afflict humankind, also applies to London's black clouds of smoke and pollution. Thought to cause deadly respiratory illness as well as corrosion to buildings such as St. Paul's cathedral, London's coal smoke problem reeked of satanic meddling.[25] Seventeenth-century religious texts reinforce the overlap between Satan's scriptural association with atmospheric evil and the infernal symbolism of coal and tobacco smoke. Their comparisons between London's air pollution and hell betray fears that Christians may be spiritually corrupted by the satanic power of coal smoke. Take, for example, *The exceeding riches of grace*, a spiritual biography published by Henry Jessey in 1647. Transformed by God's grace from a tormented young woman to a counselor of afflicted brethren, Sarah Wight, the purported subject of the biography, hears the testimony of a troubled widow who recounts a terrifying possession: "[O]ne morning, after I was awake I thought, the roome was full of smoake, and suddenly a fire went in at my mouth and went downe hot into my belly, and there it went flutter, flutter. Then (said the *woman*) I suddenly flew out of my bed, into the midst of the roome, and a voice said within mee, to my heart, *Thou are damn'd, damn'd*. I felt the smell of brimstone. Thus it began, and I thought the house was full of Devils."[26] In addition to indicating that demons are in the room, smoke seems to convey fire into the woman's belly and the smell of brimstone to her senses. Her experience echoes local conditions in London as well as scripture. An early synonym for sulfur, the word "brimstone," for instance, equates the smoke in the room with the highly sulfuric emissions of sea coal. Additionally, "brimstone" is the preferred English translation of the word *gophrith* (גָּפְרִית) from the Hebrew Bible. Possibly connected with the "gopher-wood" of Genesis 6:14, *gophrith* may have initially signified resin or sap from that tree.[27] Employing a symbol for flame and suffering, as in the verse from Deuteronomy "the whole land thereof is brimstone, and salt, and burning," the possession scene thus recalls the cursed wastelands of the Old Testament, divine wrath, and hell (KJV 29:23).

The widow's account of her possession elicits some customary guidance: do not despair for Christ has defeated the enemy. But her counselor, Sarah, also offers words of commiseration, confessing that she personally experienced a similar sense of damnation: "If all the fire and brimstone in London, and all the pitch and tarre, should all be in one fire, and I walking in the midst of that fire; this was my condition."[28] Eerily foreshadowing the Great Fire that two decades later tore through London, this hypothetical scenario is disturbing not because it is wildly farfetched, but rather, because it specifies the infernal fuels that surround them in the city. Picking up on the widow's reference to "brimstone" and echoing its relationship to the Hebrew *gophrith*, which denotes thick resin or pitch, the counselor's admission relates the demonic smoke in the widow's house to pollution on a much grander scale—the burning of London's sea coal, wood, and bitumen, or other inflammables. This apocalyptic vision of the combustion of London's fuel in one fell swoop thus carries an echo of the city's daily but nevertheless problematic consumption of coal and associates the intolerable climate produced by the household fires to the moral condition and torment of individual souls. Sarah's candid words of commiseration consist of a sympathetic, if not fatalistic illustration of the moral hazards of inhabiting a coal-powered city. No single person—no one but Christ—could possibly rid the place of this bane.

The "dedicated studies of the Devil," which began to appear in the late sixteenth century and continued to be published through the seventeenth century (roughly the same interval in which air pollution levels rapidly increased in London) portray Satan as the agent of a variety of climate ills.[29] As the author of *Of our communion and warre with angels* (1646) writes of the demons, "they are in the ayre and the world, where also they are Princes, they have the advantage of the place, and powers is also theirs."[30] The oft-republished commentary on the sixth chapter of St. Paul's Letter to the Ephesians by the popular Puritan preacher William Gouge, *The vvhole-armor of God or The spirituall furniture which God hath prouided to keepe safe euery Christian souldier from all the assaults of Satan* (1616), also suggests that the atmosphere affords Satan a fearful measure of influence over environmental phenomena and organisms.[31] Gouge provides a vivid and extensive list of Satan's capabilities in line with orthodox doctrine, for example, to "violently move the ayre, and cause tempests and stormes ... inflame the ayre, and cause thunder and lightning; yea, and extraordinary fire to fall downe" as well as "enter into" human and animal bodies to "grievously vexe and torment them, and inflict sore diseases upon them."[32] But for the armor provided by God, Gouge avers, we would be greatly outmatched by demonic enemies who are "above us, over our heads, everywhere round about us, and so still ready to annoy us," all the more so because "we fight with them in the aire, which is their Kingdome."³³

The frontispiece of the second edition of *The vvhole-armor of God* (1619), underscores Satan's territorial claims to the power of the air by associating him with smoke (see fig. 9). Representing two opposite types—the steadfast Christian and the apostate—the emperors Constantine and Julian are depicted on either side of a portico, both dressed for war. But only Constantine carries a shield, breastplate, and sword inscribed with the words "Faith," "Righteousness," and "Truth," indicating that he possesses God-given weapons with which to fend off Satan. In a group seated at the base of the illustration under a globe labeled "Faileth," a hooved devil wearing Jesuit robes clasps hands with a bare-breasted woman who cradles a goat in her lap. This triad represents "all the assaults of Satan," and their accouterments indicate the instruments through which these assaults occur: both the devil and the woman are pictured smoking tobacco pipes, and the devil holds a cross and a rifle. That the Puritans equated the Roman church with Satan explains the demon's religious garb, which probably also alludes to the Jesuit involvement in the Gunpowder Plot of 1605. The fact that gunpowder and tobacco were considered inventions of the devil accounts for the presence of the rifle and the pipes.³⁴ Employing a synthetic powder that, as some contemporaries believed, imitates a naturally occurring meteorological process, moreover, the gun portrays the devil in a weather-making capacity, as a manipulator of clouds and smoke. These smoke-related instruments associate the demonic agency shown at the base of the illustration with the billows or plumes in the sky wherein symbols of the soldiers' "rewards" are displayed: a crown of laurels for the faithful Christian, and for the apostate, a traitor's noose and hellfire falling from the smoke above. Hearkening to the lies of a priest or smoking a suspicious herb are discrete instances of the more general injury to which mere exposure to the corrupted atmosphere makes one vulnerable.

A small but significant detail above the woman's head creates a further point of connection between the demonic smoke at the base of the illustration and the more dispersed environmental haze in the upper portion of the picture. The puffs emerging from the woman's pipe waft upwards and curl around the sails of a miniature windmill. This icon possibly alludes to the wind-powered pumps used to drain England's Fens, which were often deeply resented by locals for destroying the marshes on which their livelihoods depended.³⁵ The image depicts the air that powers the windmill as corrupted by satanic figures, a devil and his fleshly consort whose rich attire suggests the greed of the projectors (like Meercraft in Jonson's *The Devil is an Ass*) who managed and profited from the drainage projects.³⁶ This remarkable detail suggests that,

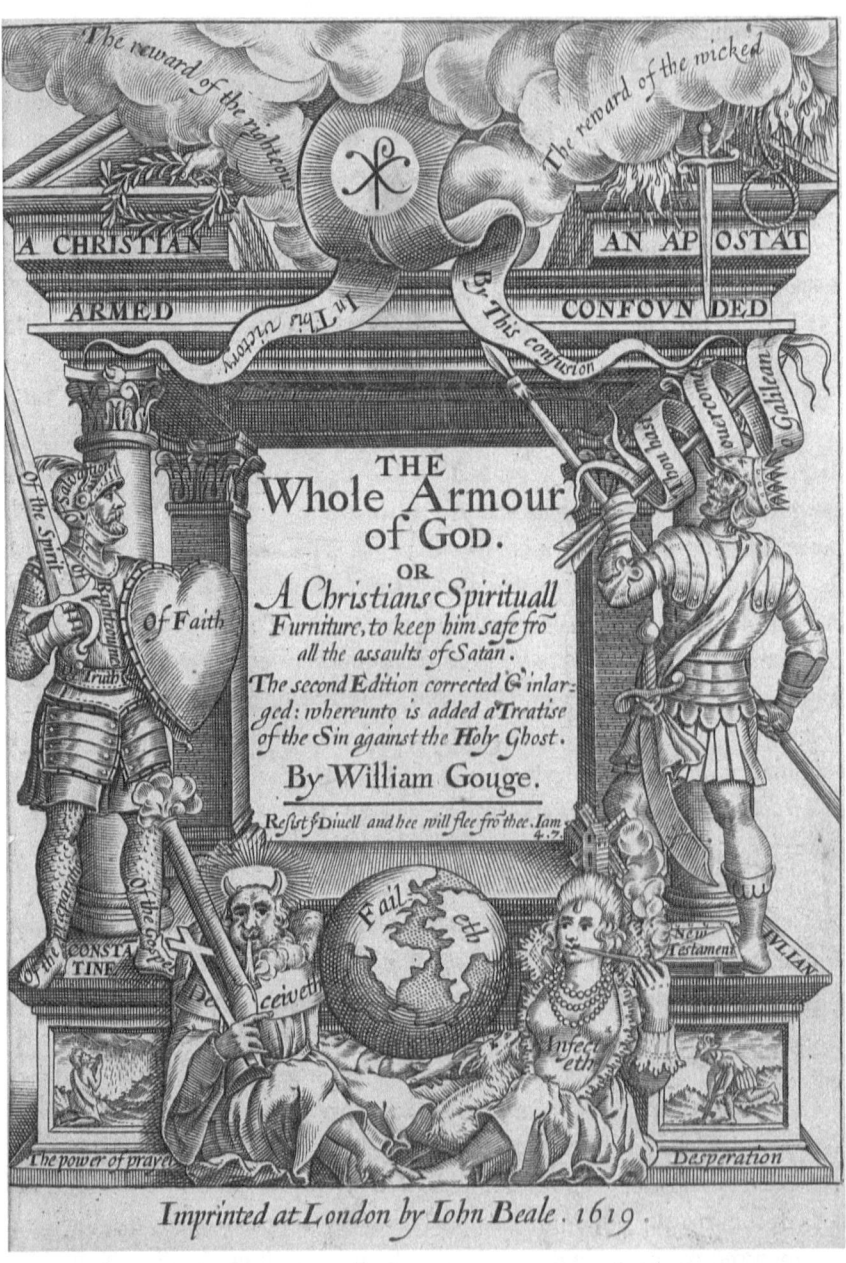

Figure 9. William Gouge (1578–1653), *The vvhole-armor of God* [. . .], 2nd ed. (London, 1619), engraved title page. (Used by permission of the Folger Shakespeare Library, call no. STC 12123)

already in this preindustrial era, smoke symbolized the social and ecological costs of extraction and unsustainable land development, and characterized the sinister, monied interests that lay behind these efforts. In illustrating subtle ties between environmental destruction and the vice associated with urban centers of consumption, the smoke depicted in Gouge's frontispiece resembles, in some respects, that of Dickens's *Hard Times,* which, as Jesse Oak Taylor argues, renders "legible" the ecological and social impact of nineteenth-century consumerism.[37] The perception of smoke in the early modern era, however, differs markedly from Dickens's view nearly two centuries later in that it presupposes Satan's actual participation in and advancement of human-made air pollution and implies that ecological disasters deserve to be condemned absolutely, on religious and moral grounds.

Given the scriptural underpinnings of Gouge's book, the illustration on its frontispiece of the Christian's agon with powers of atmospheric and environmental corruption directly pertains to Milton's *Paradise Regained*. *The vvhole-armor of God* is one of dozens of seventeenth-century primers on the subject of resisting Satan that are also, not coincidentally, commentaries on a passage of St. Paul's Epistle to the Ephesians, which exhorts the reader to "[p]ut on the whole armour of God, that ye may be able to stand against the wiles of the devill" (6:11 KJV).[38] In the epistle, this verse is followed by a detailed description of the ontology and strength of the enemy: "For wee wrestle not against flesh and blood, but against principalities, against powers, against the rulers of the darknes of this world, against spirituall wickedness in high places" (Eph. 6:12). Further elaborating the call to put on God's "whole armour," Paul's letter goes on to articulate the goal of the struggle and describe the metaphorical garments necessary for thwarting spiritual evil: "Wherfore take unto you the whole armour of God, that yee may be able to withstand in the evill day, and having done all, to stand. Stand therefore, having your loynes girt about with trueth, and having on the breast-plate of righteousnesse" (Eph. 6:13–14).

Neil Forsyth points out that the theme of standing in *Paradise Regained,* especially in the climactic pinnacle test, derives from these verses.[39] The parallels between Milton's poem and the scriptural passage are striking, for example, both repeat the word "stand" and share the idea that standing—after "having done all"—is the final step and also the result of overcoming the devil's temptations.[40] Forsyth's insight that Ephesians 6:10–14 is a main source for the final episode of *Paradise Regained*—though it implies the fundamental kinship between the two texts—emphasizes only a single part of a larger web of correspondences between this key Pauline text and Milton's poetry, especially his final masterpieces. In addition to physicalizing the epistle's language of standing and withstanding, Milton adopts its wider assumption that evil powers dwell ἐν τοῖς ἐπουρανίοις—"in high places" or "in the heavenlies"—and that godly armor is necessary to deter evil assaults from this cosmic zone.

Biblical scholars have long debated the meaning of "in the heavenlies," which is mentioned five times in Ephesians and found nowhere else in scripture.[41] The use of "in the heavenlies" is puzzling partly because wicked spirits are said to maintain authority there (Eph. 6:12) even though God raises Christ up with his believers into the same heavenly region (Eph. 2:6) and sets Christ at his right hand above all governors of this world (Eph. 1:20–21). Divine power overlaps with evil power "in the heavenlies," Arnold suggests, because the audience of the epistle feared demonic spirits of the air and needed assurance that Christ's superior position in the heavens liberates them from the demons' "dominating control."[42] The Church Fathers—Basil, Origen, and Jerome—gloss ἐν τοῖς ἐπουρανίοις (in the heavenlies) simply as "the sky" or "the air," providing a clear lexical precedent for locating the scene of spiritual combat with the Adversary in the atmosphere.[43] Moreover, the "spiritual wickedness" against which 6:12 urges the faithful to "wrestle," recalls the head of the demonic spirits mentioned earlier in the letter—the now familiar "prince of the power of the air" (Eph. 2:2). The epistle's assumption that the atmosphere is a demonic locale explains why its sixth chapter prescribes *standing* in God's armor, instead of, for example, charging forward. Stillness and spiritual preparation may be the only reasonable defenses against an atmospheric enemy who literally has us surrounded.

If Ephesians 6 is a source text for Milton, then Christ's struggle against a noisome, malignant climate is modeled on the scripture's idea that humanity fights with evil powers of the air. Since Milton's contemporaries explicitly identified Satan with urban sources of smoke (tobacco, gunpowder, and sea coal) perceptions of seventeenth-century London's polluted atmosphere also shape the brief epic's morally and physically antagonistic climate. *Paradise Regained* asserts the eschatological primacy of reclaiming the atmosphere from the agents of its corruption. Perfectly confuting Satan's denigrations of the air and reshaping it into truth epitomizes Milton's idea of heroism and foreshadows the final reclamation of the fallen territory of the air for humankind.

The contest between Satan and Christ in *Paradise Regained* has been called a "battle of the wits"—an intellectual and spiritual duel that commences on an even playing field with both contestants possessing an approximately equal measure of "fallible, imperfect knowledge."[44] This evenhanded, intellectual framing of the battle, however, diminishes the physical colossus of Christ's foe whose authority extends even into the elemental foundations of the world. At the beginning of the poem, God promises to "expose" Christ to Satan, signaling that the enemy's evil, like a pestilential fume or a raging storm, is outside, around, or above (1.143). To make matters worse, in the wilderness Christ lacks provisions—not even the shed that at his birth was found, as Mary recalls, "to shelter him or me / From the bleak air" (2.73). Christ's heroism, though based in intellect and faith, also includes physical toughness. He literally weathers Satan's assaults when the storm strikes him in book 4, but also in the first and second temptations. Christ's challenge is to withstand the influence that Satan wields as a meteorological power throughout his sprawling empire. By physically elevating Christ onto the pinnacle of Herod's temple and into the atmosphere, his own stronghold, Satan forces the question implicit in the other temptations: can and will the Son of God extinguish the power that Satan enjoys in the air?

In *Paradise Lost* Milton eschews the martial or chivalric virtues typically celebrated by epics and their heroes, preferring instead "patience and heroic martyrdom" (9.32). Scholars have argued that Milton models his notion of heroic virtue on Aristotle's *Nicomachean Ethics*, the locus classicus of the Renaissance tradition of the "magnanimous man."[45] But another probable model exists in a passage discussed earlier from *The City of God* that pointedly juxtaposes the Christian martyrs with pagan heroes. Since Augustine defines Christian heroic virtue in distinction to the Greco-Roman ideal, it is particularly relevant to Milton's "brief biblical epic," which adapts a classical genre in a biblical or Christian vein.[46] Augustine's portrayal of defenders of the faith anticipates Milton's notion of "better fortitude / Of patience and heroic martyrdom" (9.31–32): "[T]he city of God holds her *martyred* citizens in greater honour and higher repute in such measure as they fight with *fortitude* against the sin of denying their religion, even to the shedding of their blood."[47] The greater the fortitude he or she displays, the more highly esteemed the martyr. Next, the text raises the possibility that a martyr is a type of hero: "If the Church's traditional canon of style permitted such a term, we might more neatly dub [the martyrs] 'our heroes'."[48] Sharing the air as a dwelling place with evil demons and taking their name (hero) from its deity (Hera), however, these pagan paragons of virtue differ in important respects from Augustine's "heroes." A crucial difference is that they placate the powers of the air and their leader. To that end, as Augustine observes, Helenus advises Aeneas "[t]o Juno gladly chant your prayers and overbear / That mighty queen with suppliant gifts."[49] From the church's perspective, this is a heroism of cowardice and complicity: "This is not the way of true and truly holy religion; not thus do our martyrs overcome Juno, which is to say the powers of the air that envy the valour of the god-fearing. Our heroes, if usage permitted us so to call them, do not in the least resort to gifts to overbear Hera, but to valour that comes from God. Surely it was better that Scipio should win his surname Africanus by conquering Africa valorously, than by appeasing the enemy with gifts and persuading them to have mercy."[50] Heroic virtue as construed by the church thus entails overcoming the powers of the air without also capitulating to them. The final sentence argues that according to the convention that gave Scipio the surname "Africanus," those who vanquish Hera ought to be known as heroes.[51] It also implies that resisting the powers of the air, much like a Roman defeating Africans on their own turf, represents a meaningful inroad into enemy territory. Molding the hero of his short epic into the warlike martyr that Augustine describes, one who uses God's valor to undermine the powers of the air, Milton depicts Christ's temptations as a spiritual battle with the prince of the air and ultimately an epic judgment on the devil's dominion over the atmosphere.

In *Paradise Lost,* constraint of the devil's atmospheric power is presented as a primary reason for the Incarnation. In his judgment of the serpent, the Son foretells that he will watch the prince of the air "fall like lightning down from heaven," and parade him as "captive through the air," a prisoner in his own ill-gotten realm.[52] In the sequel to *Paradise Lost,* the Son meets his adversary before these victories occur and while the demons still "[r]ule in the clouds" (4.619). Satan openly glories in the titles and other advantages the devils derive from their place. He venerates his fellow demons by calling them "ancient powers of air and this wide world" (1.44). Having fulfilled a promise to hells' citizens to resettle them on earth and in the atmosphere, Satan touts the privileges he thereby procured, such as the "liberty to round this globe of earth, / Or range in the air" (1.365–66). The anointing of God's Son, however, means that all could be "infringed, our freedom and our being / In this fair empire won of earth and air" (1.62–63).

Reminding his followers of the freedoms they enjoy in the air and the vast superiority of Earth's climate to that of hell, Satan uses rhetoric of self-aggrandizement (which the Son punctures at 1.409–20). There is a grain of truth to Satan's boasts. As biblical commentators recognized, the devils' preeminence in the atmosphere represents a real, but not insurmountable, advantage over humans. Like "a few men on a hill, or on high walles and Towers," Gouge says of the powers of air, "[t]hey can espy all things that we doe" and can inflict "much mischief to a great Army in a low valley beneath them."[53] The Son of God appears as earth-bound as other humans, and hence susceptible to atmospheric manipulation.[54] Satan exploits this advantage, for example, when he attempts to dazzle Christ with visions of distant kingdoms. Recalling the passage from *Paradise Lost* that explains the serpent's vocalization through alternatives, either by "tongue / Organic" or "impulse of vocal air" (9.529–30), the brief epic proposes at least three techniques that Satan may have used to give Christ a close-up look at Rome despite mountains that blocked the view. Of the trio of alternatives put forth,

> strange parallax or optic skill
> Of vision multiplied through air, or glass
> Of telescope,
>
> (4.40–42)

the second involves manipulating the air. The "airy microscope" that Satan brags about a few lines later, which can reveal the gilded exteriors *and* interiors of the cityscape, combines the second and third alternatives: pneumatic optics and visual instrumentation (4.57).[55] As in his addresses to Eve, Satan uses both mechanical instrumentation and pneumatic influence in his temptations of Jesus. In Milton's brief epic, however, Satan increasingly lays stress on another, closely related arm of his power—control over the elements.

We can see this subtle shift in the archfiend's use of denominations. In their first meeting, Christ sharply denies the significance of Satan's place in the heavens, and he returns to his council in the air deflated, "without sign of boast, or sign of joy" (2.119). He greets them as

> Princes, heaven's ancient sons, ethereal thrones,
> Demonian spirits now, from the element
> Each of his reign allotted, rightlier called,
> Powers of fire, air, water, and earth beneath.
>
> (2.121–24)

With Christ's statement that Satan is "never more in hell than when in heaven" still ringing in his ears, Satan retires the demons' heavenly titles and replaces them with names derived from lower cosmological strata (1.420). He calls the demons powers of fire, water, and earth, in addition to air. Satan's emphasis on the demonic power distributed throughout the environment reflects his initial misunderstanding of the type of kingdom Christ is to inherit. Hoping to illustrate the necessity of accepting his lordship, Satan alludes to the tribute he receives from all earthly rulers and nations including his own "Tetrarchs of fire, air, flood, and on the earth" (4.201). A king cannot function without the agreement of the elements, or so Satan assumes, and these "mild seats" belong to the demons (2.125).

The demonic power of the elements depicted in *Paradise Regained* is familiar from Milton's *Il Penseroso*, where the contemplative man fantasizes about consulting Hermes Trismestigus, or Plato, to learn

> of those demons that are found
> In fire, air, flood, or under ground,
> Whose power hath a true consent
> With planet, or with element.
> (93–96)

Milton himself might have learned about the Hermetic philosophy by reading Patristic authors, which his works frequently cite.[56] According to writings attributed to Hermes, the gods and the stars influence affairs on earth through demonic agents who dwell in each layer of the universe.[57] Although the origin of this theory differs from that of the doctrine of the prince of the power of the air, which associates demons mainly with air rather than all of the elements, the two positions are nonetheless compatible with each other. As we saw in chapter 3, Christian theologians and natural philosophers envisioned the demons as agents of divine retribution with power over a variety of meteorological phenomena. Similarly, in the *Corpus Hermeticum* 16, the gods enjoin the demons to punish mortals "through torrents, hurricanes, thunderstorms, fiery alterations and earthquakes."[58] Acknowledging that bad angels "wander all over the earth, the air, and even heaven" Milton's demonology encompasses both the ancient Hermetic and Christian traditions, locating evil powers in and above the air, and below on earth. With the scriptural warrant of Galatians 4:3, for instance, which portrays humanity as children in "bondage under the elements of the world," Milton conceives of satanic corruption as extending from the air into other rudiments of the universe: fire, water, and earth.[59]

ELEMENTAL DEMONS AND THE FALSE FUEL CHOICE

Along with the atmospheric power that has always served Satan's malicious plots, he conducts the temptations by harnessing other corrupted elements in the physical environment. In his first disguise as "an aged man in rural weeds," Satan appears to be practicing prudent husbandry,

> the quest of some stray ewe,
> Or withered sticks to gather; which might serve
> Against a winter's day when winds blow keen,
> To warm him wet returned from field at eve.
> (1.314–19)

But in reality these activities represent the misallocation of natural resources. The passage initially analogizes the aged man to the good shepherd in the parable of the lost sheep (Luke 15:3–7; Matthew 18:12–14), but the swift transition in the following line casts doubt on the man's motivations for finding the "stray ewe." As readers have noticed, the allusion to gathering "withered sticks" recalls two foreboding passages from scripture: Numbers 15:32–36, in which a man is stoned to death for gathering sticks in the wilderness on the Sabbath day, and the parable of the vine wherein Jesus compares those who abide in him to branches that bear fruit, and others to withered branches that men gather up for burning (John 15:6).[60] In light of these texts, the old man's seeming occupations appear either unlawful or sinister. Whether he recovers a wayward sheep or gathers withered branches, neither is done innocently.

The less often remarked part of the description is the imbedded allusion to Satan's office as the prince of the air and the power he will use later in the epic to produce a storm. Although Satan usually inflicts stormy weather on others, in this caricature, *he* is the person who seeks relief from winds that "blow keen." The means by which he seeks to purchase relief are significant. Searching the ground for fuel to make fire, he hopes to inure himself to other elements: the downpour of "a winter's day." One ironic effect of the implied sequence of events—a winter's dousing followed by a warming fire—is to foreshadow Satan's defeat by Christ and eventual fiery punishment in hell. Another is to lay bare the unsustainable economy and ecology of Satan's elemental power. Attempting to gird himself against atmospheric corruption by burning earthly fuels—sticks symbolic of lost human souls—is self-defeating and futile. It doubles down on a climate problem of Satan's own making. We see this in the description's lack of closure and implication of disappointment. It is doubtful that the man collects any firewood at all, and at some point he will return home cold and "wet" from the field.

The fact that Milton imagines Satan as having the look of someone searching for firewood, moreover, evokes the prehistory of early modern England's fuel crisis and eventual turn to coal. During the century before Milton's birth, wood went from being the primary way to heat the home to an increasingly unaffordable option. Along with other commodity markets, wood supplies showed signs of strain beginning in the sixteenth century as the population nearly doubled from the 1530s to the 1650s. "The inevitable consequence of a strong rise in the demand for wood in excess of any increase in production, was scarcity and inflation."[61] Although the effect on England's woodlands was by no means uniform, supplies to a growing number of localities fell grossly short of the demand. Because of simultaneous food shortages, the lightly wooded areas around towns and cities were often converted to arable farmland, cutting off local populations from easily accessible firewood.[62] Situated on the eastern coast in the least wooded part of England, London shipped and carted in wood supplies from distant forests in Essex, Kent, and Suffolk.[63] But such measures were ultimately not sufficient for the country's largest fuel market. Europe was in the midst of the Little Ice Age and the poor suffered in the freezing weather. With the price of wood continuing to skyrocket throughout the early seventeenth century, London made the switch to the cheaper alternative of coal.

The precariousness of the fuel market that preceded England's coal-based economy illuminates Satan's depiction as a wood collector and his response to the limited resources of the wilderness. Communities affected by the fuel crisis throughout sixteenth-century England began to police wood supplies, imposing severe penalties such as corporal punishment and expulsion on those who pilfered wood or engaged in "hedgestealing."[64] The criminalization of gathering wood in early modern England sheds light on the connection between Satan's initial disguise and the biblical story from Numbers of the man who is put to death for gathering sticks on the Sabbath. It is not immediately clear why Milton, or the Israelites for that matter, considered the man's act a desecration of the Sabbath.[65] Exodus 31:14–15 prohibits working on the seventh day under penalty of death—as does Exodus 35:2. But it is unclear whether the man's actions, "gathering sticks," to use the phrase from the King James Version, constitute work. Given the strict limitations on fuel in their own era early modern Christians may have been prone to read the stoning story in light of another relevant declaration from Exodus that "ye shall kindle no fire throughout your habitations upon the sabbath day" (35:3). Indeed, Milton focuses on how the man in Numbers 15:32–36 potentially violates this anti-hearth-fire law, since the aged man in *Paradise Regained* appears to gather wood in order to "warm him" at the end of a cold day. Satan's initial disguise thus portrays a person whose misuse of earthly fuels poses a moral (and mortal) threat to his community, as a pilferer might jeopardize townsfolk who have a limited supply of wood.

Throughout his first and second encounters with Jesus, Satan invokes concerns about austerity and expense, ideas which dominated the fuel crisis before the shift to coal and later resurfaced during the seventeenth century, particularly when England was at war and London's coal supply was threatened.[66] Though there are obvious contrasts between the "desert wild" of *Paradise Regained* and London, they share a common remoteness from resource-rich areas that might supply their inhabitants' needs (1.193). Firewood could travel tens of miles from the surrounding country in order to reach Londoners,[67] and coal was brought a much longer distance by sea route from Newcastle. "So far from path or road of men, who pass / In troop or caravan," the wilderness in *Paradise Regained* is also a great distance from supply routes (1.322–33). According to Satan, those who dwell there are forced to "[l]ive on tough roots and stubs" (1.339). This phrase characterizes the local diet as extremely limited and crude. "Roots" probably refers to simple vegetables, such as turnips or potatoes. But "stubs" has a slightly more ambiguous meaning as it relates both to agriculture and fuel. In the agricultural sense, "stubble" is what is left in a field after the upper part of the crop has been reaped.[68] "Stub" also refers to the stump of a tree, the lowest portion of a trunk, or the "stubwood" found in hedgerows—an important source of firewood for early modern people.[69] In a contemporary sermon on the topic of marriage "stubble" is used ambiguously—either as foodstuff or fuel—to illustrate a potentially inflammatory point of contention that still can be easily defused: "In this case the caution is to subtract fuel from the sudden flame; for stubble though it be quickly kindled, yet it is as soon extinguished if it be not blown by a pertinacious breath, or fed with new materials."[70] Yet Satan's breath is more than pertinacious; it blows unrelentingly on whatever kindling will serve his argument. By mentioning stubs along with roots—useful dregs of England's farmlands and woodlands—Satan suggests the inexorable connection between fuel and food. During times of shortage both were necessary for survival, but because of the nature of pre-modern economies, each came at the expense of the other.

In some respects, fuel and food in an "organic economy," the kind that England possessed prior to the seventeenth century, differed very little.[71] The difference in edibility between "root" and "stub," for example, is a matter of degree, as is the difference in their potential usefulness as fuel. A root is more suitable for eating, and a stub, slightly better for burning. Both, at their most basic, are forms of vegetation, which is a problem since they must occupy a finite amount of land. Growing in similar places, crops and coppiced wood were often in competition for valuable acreage.[72] In a coal-based economy, however, the difference between fuel and food is much more stark. Coal looks and feels like a rock, hence its association with brimstone. More importantly, coal is harvested from beneath the ground and thus does not encroach on farmland. By relaxing demands on the land, the shift to coal at the outset of the seventeenth century contributed to a sharp rise in agricultural productivity over the next two centuries.[73]

Setting aside the devastating environmental implications of adopting coal, this breakthrough afforded the economic and technological wherewithal for England's eventual rise as an industrial powerhouse. Yet in Milton's day these gains were still far off, and even the earliest putative benefits of switching to coal—a steady stream of affordable thermal energy and a greater abundance of food—did not always materialize or satisfy the needs of early modern Londoners. In the first decades of the new fossil fuel era, and despite the actual antithesis between coal and nourishing food, the analogy between fuel dearth and famine was frequently invoked. Officials and projectors likened the dependence of London's poor on coal to their need for food.[74] Whereas such comparisons did not mean "that early modern Londoners were likely to confuse bread with coal," as William Cavert suggests, "there was real conceptual overlap between these distinct types of necessities."[75]

Satan's complaint about living with scarcity implies that conversion of fuel to food is necessary for survival. Meddling with sticks or stubs, however, is not what he has in mind. He now speaks as a coal prospector:

> But if thou be the Son of God, command
> That out of these hard stones be made thee bread;
> So shalt thou save thyself and us relieve
> With food, whereof we wretched seldom taste.
> (1.342–45)

"Stone" is the word the devil uses in the biblical story (Luke 4:3–4). In light of the poem's previous allusions to wood fuels and a later reference to Elijah waking up to "his supper on the coals prepared," here, stone serves as a metonym for coal (2.273). There is precedent for this linguistic substitution. On first witnessing the use of coal during the mid-sixteenth century, for example, a traveler marveled that the poor in Scotland were happy to accept "stones" (*acceptis lapidibus*) for alms, which they burned because almost no wood was to be had in that country.[76] Fossil fuel, however, is a far cry from "bread" especially because it has to be mined, processed, and shipped before it resembles something like a household fuel that can be used to make bread.[77] Satan asks the Son of God to elide all of these steps, turning minerals directly into a meal.

One of the objectives of Satan's question is to test whether Jesus can and will perform a miracle. But as we have already seen, the feat that he proposes is something that early modern Londoners sought in reality. Drawing parallels between coal and food—sometimes on behalf of the poor, sometimes for the sake of other interests—pamphleteers and politicians commanded that Newcastle's stones be turned into London's bread, cheaply and expeditiously.[78] Others accused suppliers of holding the market hostage. In a petition addressed to Oliver Cromwell, for example, Ralph Gardiner argues that the poor "have found it as hard a matter to fortifie themselves against cold, as against hunger" because of the Corporation of Newcastle's "tyrannical" monopoly over the coal trade.[79] The Crown granted a charter to the city in 1600, giving the exclusive right to ship coal as well as control over the municipal government to the Company of Hostmen, a medieval guild that over the centuries had morphed into a powerful coal corporation.[80] Gardiner accuses these magistrates, supposedly representatives of God's "Power and Justice," of corrupting their offices "into the Image of Gods enemy, Satan, whom herein they resemble, and become after a sort wickednesses in high places, as the Devils are."[81] Explicitly comparing Newcastle's "Lords of Coal"[82] with Satan and the devils in "high places" of Ephesians 6:12, Gardiner draws on the satanic associations of coal and atmospheric pollution. Drawing a similar connection, Milton depicts the satanic prince of the air as dealing in the "combustible / And fueled entrails" of the earth—minerals that feed the atmospheric corruption and meteorological terror of his earthly reign. In tempting Jesus to turn stones into bread, Satan seeks to trick him into betraying his faith, but also to "behold thy godlike deeds"—that is, to probe the extent of the Son's power (1.386). The sign of power that he requests is one that a coal king, or fossil-fuel dependent England, would be apt to recognize: the ability to make provision from mineral resources.

The Son sees through the tempter's feigned concern for feeding the "wretched." Satan exploits environmental resources to harm humankind, not nourish it. Moreover, there is no elemental fuel or food that can give him succor. "Lying," as Christ informs Satan, "is thy sustenance, thy food" (1.429). Thus, in the next part of the temptation, Satan drops the pretense of seeking only the bare necessities for himself or others. Shedding his beggar's costume and air of humility, Satan's "regal" banquet belies the modesty of his previous plea for bread from local stones. *That* appeal—for the merciful and easy conversion of fuel into food—is perhaps as mythical as the modern notion of "clean coal." Those who were acquainted with London's coal market and its elaborate supply chain knew that fuel reached the consumer neither swiftly nor uncorrupted. As Gardiner fumes in his anti-coal-monopoly tract, at every turn of the labyrinthine system, producers and shippers took their cut, ratcheting up the price. Moreover, tainted coal was lumped into the final product.[83] The mounting moral and environmental costs of Satan's feast, as we shall see, demonstrate that no earthly bounty, whether fuel for the home or meats for the table, can be given freely by human or demonian courier.

SATANIC HOSPITALITY AND THE "PEELING" OF THE EARTH

Satan's table is set, not surprisingly, in a wood ensconced in exotic and dreamlike atmospheric effects. "Resounding loud" with music that Jesus listens to from above, the "pleasant grove" echoes with the "chant of tuneful birds" (2.289–90). Yet, no living birds are seen inside of this "high-roofed" wood (2.293). Instead, Satan's "chosen band / Of spirits," musicians who practice the great serpent-player's art, dissemble the natural sounding birdsong that emanates from the grove with "chiming strings, or charming pipes" (2.236–37, 363). As in Paradise, the atmosphere of the wood is simultaneously aural and meteorological. The "harmonious airs" of the grove are also "winds / Of gentlest gale" (2.363–64). But the atmosphere here has been corrupted. Endowed with bird-like "soft wings" and "Flora's earliest smells," the breezes intermix Nature's wares with "Arabian odours," merchandise from the spice trade (2.364–65).[84] The winds of Eden in *Paradise Lost* carry "[n]ative perfumes," but this wilderness is "[o]f all things destitute" and thus everything in the enclave, even the spiced winds, are spirited from elsewhere (*PL* 4.158; *PR* 2.305).

The grove's pseudo-Paradisal atmosphere infuses the feast laid before Christ. The table resembles Edenic cuisine in its copiousness but differs in that the main offering is meat. The question is whether this food may be accepted. Satan hopes that Christ will be induced to sin by acting on an impulse "which only seems to satisfy / Lawful desires of nature" (2.229–30). Alluding to the scriptures regarding Christ's title as the world's heir and man's dominance over the beasts, Satan argues that Christ has "right to all created things" (Heb. 1:2; Gen. 1:26; 2.324). Jesus has a more nuanced understanding of his relationship to creation. During another biblical meal, the Last Supper, Christ portrays the bread that he breaks and gives to his disciples as his body. The food lying on the wilderness banquet table travesties the Eucharist, but the parallel is nevertheless unmistakable. His reflections about his hunger reinforce this parallel as they reveal that Christ thinks of himself as connected to Nature and subject to its needs:

> now I feel I hunger, which declares,
> Nature hath need of what she asks; yet God
> Can satisfy that need some other way,
> Though hunger still remain: so it remain
> Without this body's wasting, I content me.
> (2.252–56)

It is tempting to gloss these lines as if "Nature" stands solely for Christ's *human* nature: Christ speaks of trusting God to satisfy his human needs so long as his body does not deteriorate. This is a valid and, perhaps, the dominant interpretation of the lines; but it is worth considering whether they also signify Christ's view of a more encompassing idea of Nature.

Christ's view comes into sharper focus in juxtaposition to Satan's portrayal of Nature. Satan tries to distinguish and hierarchize the concepts of human nature and the natural world. He wants to depict Christ's hunger as a "lawful desire of nature," yet in order to persuade him to satisfy this natural desire, ironically, Satan must recommend the degradation of Nature-as-creation. Thus, in presenting the banquet table to Christ, he directs him to

> behold
> Nature ashamed, or better to express,
> Troubled that thou shouldst hunger, hath purveyed
> From all the elements her choicest store
> To treat thee as beseems, and as her Lord
> With honour, only deign to sit and eat.
> (2.331–36)

Though the speech ends with the idea of honor, it begins with the opposite concept—a characterization of the delicacies offered to Jesus as "Nature *ashamed*."[85] Satan quickly corrects course in the next line, applying the softer adjective "troubled," and goes on to pretend that Christ's hunger is the cause of Nature's discomfort. But his original, uncensored description of the meal is closer to the truth. When demonic "[p]owers of fire, air, water, and earth beneath" conjure from "all the elements" a lavish feast to tempt the Lord, there is cause for Nature's shame. In the *Nativity Ode*, we find Nature in a similar state of embarrassment. She is riddled with lying oracles and demons and anxious to hide "her naked shame" from the Savior. In the above passage, however, instead of lamenting Nature's corruption, Satan brags about the despoiled environment and argues that Christ has the right to enjoy the plunder.

Instead of splitting off and privileging human wants above all else, Christ's view that "Nature hath need of what she asks" expresses that need is universal. *All* of Nature hungers, not just humanity. This notion clashes especially with the satanic fiction that bounty and rarity can coexist. The scarcity he warned of as lowly wood collector is far from a concern at the banquet—the table is "richly spread" and set with "dishes piled"—nor is the food "simple," like the "crude apple that diverted Eve!" (2.340–41, 348–49). Purveying outrageously exotic cuisine is not merely an expression of Satan's assumed nobility (in this episode he dresses as one bred in "court, or palace"), it is integral to his demonological identity (2.300). In Thomas Adams' book of sermons, *The Devil's Banket*, for example, anomaly and expense are the main items on Satan's menu: "There is no man in the world keepes such hospitalitie, for hee searcheth the ayre, earth, sea, nay, the Kitchen of Hell, to fit euery palate. *Vitellius* searched farre and wide for the rarities of nature; Birdes, Beasts, Fishes of inestimable price; which yet brought in, the bodies are scorned, and onely the eye of this Bird, the tongue of that Fish is taken: that the spoyles of many might bee sacrifices to one supper. The Emperour of (the low Countries) Hell, hath delicates of stranger varitie, curiositie."[86] Glancing at the rulers of the Spanish empire, the passage compares satanic luxury to that of the glutton Emperor Vitellius.[87] Adams might have singled out other famous gourmands[88]—his point is that Satan's nicety in procuring the strangest, richest foods surpasses even the obscenest Roman tastes.

Multiple parallels between the passage from *The Devil's Banket* and Milton's banquet scene shed light on the immoral economy of Satan's environmental power in *Paradise Regained*. Adams's devil searches the elements of the world, "ayre, earth, sea," and the "Kitchen of Hell," for the best specimens. Similarly, in *Paradise Regained*, Satan's alliance with the "fire, air, flood," and the "world beneath" entitles him to cull from "the elements [Nature's] choicest store" (2.334, 4.203). Spiritual occupation of the environment ideally positions the devil to wrest precious resources from the physical world. Another parallel lies in the Roman emperor's wide-ranging appetite for rare delicacies. Assembling out of slivers of meat an Arcimboldo-like portrait, Vitellius's cuisine from "farre and wide" brings together the "eye" of one rare beast, the "tongue" of another.[89] In the process, animal "bodies are scorned" and "the spoyles of many" (meals that others might have eaten) are "sacrifice[d]" for the supper of a solitary man. Milton's description of Satan's train of meats similarly emphasizes their distant origins and hints at the violence associated with harvesting them. His table features

> beasts of chase, or fowl of game
> In pastry built, or from the spit, or boiled,
> Grisamber-steamed; all fish from sea or shore,
> Freshet, or purling brook, of shell or fin,
> And exquisitest name, for which was drained
> Pontus and Lucrine bay, and Afric coast.
> (2.342–47)

The beasts and fowl are hunted by cruel sports, "chase" and "game," and the fragrance used for steaming the meat—"Grisamber"—has its own history of violence. An ingredient of perfumes and medicines, ambergris was by one early modern estimate "twice as valuable as gold," and though its origins were not fully understood, explorers and travel writers throughout the period report the hunting of whales for the ambergris found in their bellies.[90] If slaying Leviathan for the sake of a prized intestinal stone epitomizes the wasteful butchery of satanic "hospitalitie," then the banquet's piscine dishes portray biological and environmental waste on an even grander scale. Alluding to the Roman gourmandise described by Adams, Milton's table is stocked with every kind of fish including those of "exquisitest name,"[91] which come from the Black Sea, known in antiquity as Pontus Euxinus, the "hospitable sea"; from Italy's Lucrine Lake, noted by classical authors for its mussels and oysters; and from the neighboring "Afric coast." With its own Tyrrhenian Sea exhausted by gluttony and overfishing, as Juvenal claims, Rome had to import fish from its surrounding territories.[92] Imagining a wider disaster zone than Juvenal does, Milton's word "drained" implies the depletion of even those outlying provincial waters. Given the implication that eating "exquisitest" fish decimates foreign habitats and food supplies, their presence at the banquet undermines Satan's assurances that tasting the viands "destroys life's enemy" and "life preserves" (2.372). In light of Christ's concern for "this body's wasting" as a consequence of Nature's deprivation, moreover, it is easy to see why Christ rejects Satan's ill-gotten fish and the other dishes, which entail so much bodily waste.

Affecting to provide fuel and food to human beings, Satan's program of environmental corruption rests on an insidious form of obligation. We have already seen how he uses the rhetoric of honor and entitlement to flatter Christ. This nostalgic diction, itself rooted in a culture of mutual obligation, masks a more brutal underlying view of relationships as always either mercenary or enforced. This slippage between fealty and compulsion is evident in Satan's question to Jesus:

> Owe not all creatures by just right to thee
> Duty and service, nor to stay till bid,
> But tender all their power?
>
> (2.321–27)

Duty and service are not necessarily compulsory, but Satan's use of the words "owe" and "tender" suggests that creatures should offer their power in order to pay a material debt.[93] Indicating that creatures should bring their gifts or service to God unbidden, Satan's query diverges sharply from the portrayal of worshippers who merely "stand and wait" in "Sonnet XVI." The different attitudes embodied by the desert's "wild beasts" versus Satan's "spirits of air, and woods, and springs" is telling: the former creatures stay a distance from Christ and watch with caution, and the latter surrogates appear in eye-catching array, ready "to pay / [Him] homage" (1.310–13, 2.374–76). However, Christ refutes Satan's exhortation to accept "power" from created beings: "Shall I receive by gift what of my own, / ... I can command?" (2.381–82). Satan fails to recognize that offerings are not effective with the Son of God, though they may work on a power of air like Hera or himself, who, according to Augustine and Virgil, may be appeased or overborn with "supplicant gifts." Satan admits to using this strategy of disarmament, though to no avail. The rich offerings of the banquet were intended to subdue Christ by satisfying his hunger, that which "each other creature tames" (2.406).

Even though the Son exposes Satan's offerings for what they are, "no gifts but guiles," Satan still flaunts his elemental power and the obligations that flow from it (2.391). Claiming to possess "power to give" and pretending not to act under constraint, nor seek anything in return ("of that power I bring thee voluntary"), Satan tries to portray his donation as that of a benevolent and disinterested lord (2.393–94). Yet, later in the temptation of the kingdoms Satan backtracks saying he does not "mean to give for naught" (4.161). He begins to drop his claim to artless largess almost immediately after the banquet temptation collapses. He tells Christ that he gave him "[w]hat I might have bestowed on whom I pleased" implying that Christ snubbed food that might have fed other deserving individuals (2.395). It would be a mistake, however, to picture those individuals as righteous men or the hungry poor. Explaining where he will send the delicacies next, Satan says, "[o]f these things others quickly will dispose / Whose pains have earned the far-fet spoil" (2.400–401). His claim that the happy new recipients of the food "earned" it echoes God's assertion that he chose a "perfect man, by merit called my Son, / To *earn* salvation for the sons of men" (1.166–67; emphasis added). Yet, the satanic and divine definitions of "merit" could not be more different. The henchmen Satan seeks to reward take "pains" to ingratiate themselves within the fallen world order—they work to gain influence and riches, not to liberate the human race.

Other signs point to the self-interested motives of Satan's co-partners and their implication in his environmental abuses. The "sound of harpies' wings, and talons," the acoustical trace of the invisible powers of air who take the viands away, ironically echoes the story of ravens bringing food to the prophet Elijah in the desert mentioned twice at 2.403 and 2.312–14. As the narrator notes, the ravens were "taught to abstain from what they brought," despite being "ravenous" as their name suggests (2.269). Satan's spoils on the other hand are earmarked for those who "earned" them, presumably, harpy-winged demons like those that brought the "far-fet" viands from distant lands and now take them away. The demonian ranks may receive payment for their efforts, but Christ utterly dismisses their vain "diligence" as the "toils of fools"[94]—the assiduous quest for trophies of wealth, which corrupts and encumbers man's inner virtue (2.387, 453).

Satan's duplicitous approach to governing the earth's elements—invoking scarcity on the one hand to justify calls for provision and creating scarcity on the other hand to fuel excess—inheres in the policies of the human empires over which he presides. In book 3, Satan takes Christ to the top of a mountain above Mesopotamia where he can observe "[t]he monarchies of the earth" and learn how to attain his destined throne (246). The land resembles Paradise in its apparent abundance: "[f]ertile of corn the glebe, of oil and wine, / With herds the pastures thronged, with flocks the hills" (3.246). But there are also reminders that the landscape is fallen. Pockets "here and there" of "barren desert fountainless and dry" signal that even amidst fertility lurks death and privation (3.263–64). Threatening to increase the disparity between fertile glebe and barren desert, the powerful kings of this land do little to preserve its natural riches or share them equitably. After listing off several landmarks of ancient Assyria, for example, Satan comes to the river "Choaspes, amber stream, / The drink of none but kings," echoing the legend of the "golden" water of ancient Persia reserved only for the king and his son (3.288–89).[95] According to another story that recalls the satanic labor of hauling rare delicacies into the desert, the Persian emperor Cyrus refused to drink from any river except the Choaspes and, accordingly, took wagon loads of silver vessels filled with the precious liquid wherever he went (3.284).[96] Not only do the region's rulers hoard its resources, but they also sacrifice them for the ends of war. Marching from their seat in Ctesiphon on the Scythians in far-flung Sogdiana, the Parthians carve up the intervening land, dismantling terrain that might have served as their granary or as a source of fuel:

> A multitude with spades and axes armed
> To lay hills plain, fell woods, or valleys fill,
> Or where plain was raise hill, or overlay
> With bridges rivers proud, as with a yoke.
> (3.332–34)

Unwilling to leave natural features as they were, the army's pioneers destroy whatever hills, forests, valleys, and plains lie in their path, and subdue rivers with an imperial "yoke."

Rejecting Satan's offer to broker an alliance with the Parthians or secure their surrender, Christ is made to consider the Roman Empire to the west. Like the farmlands and pastures of the Fertile Crescent, the fields of Latium are also disposed to flourish since they are

> backed with a ridge of hills
> That screened the fruits of the earth and seats of men
> From cold Septentrion blasts.
>
> (4.29–31)

Shielded as they are from northern winds, blasts that the prince of the air controls, the Romans ought to have built a society both prosperous and pious on this fertile plot. But as Christ explains, victory and wealth lulled the people into degeneracy. As the Parthians subjugate and destroy when they spread out, so the Romans

> govern ill the nations under yoke,
> Peeling their provinces, exhausted all
> By lust and rapine.
>
> (4.135–37)

Far from sheltering itself from the fallen climate, the Romans "peel[ed]" or denuded the empire of that which fed and protected it. The Romans' "sumptuous gluttonies, and gorgeous feasts," which are served on imported African wood or "Atlantic stone," recall Satan's banquet with its selection of fish brought from "Afric coast" (4.114–15). So slight is the difference between the Roman and infernal empires that they use the very same "exhausted" provinces to stock their tables.

Christ answers Satan's repeated promises to speed him to David's throne through riches, arms, and glory, by condemning these means as false and deleterious, and averring his trust in God's providence. Yet even in the face of Christ's implacable self-possession and the prospect of his own demise, Satan summons the will to flatter. Complimenting the Son on his peaceful mien, he wonders whether Christ's reign might offer him refuge from God's "ire," a "shelter and a kind of shading cool / Interposition, as a summer's cloud" (3.221–22). The irony of this vain wish is striking in light of the devastating consequences of Satan's atmospheric and elemental power acting in the world. All the glory amassed by earthly kingdoms through "conquest far and wide"—campaigns to "rob and spoil, burn, slaughter, and enslave / Peaceable nations" and "leave behind / Nothing but ruin whersoe'ver they rove"—Satan attributes to himself (3.72, 75–76, 78–79). "To me the power / Is given, and by that right I give it thee" he tells Christ (4.104–5). Despite sponsoring such ruination, the draining, felling, peeling, mining, and spoiling of the earth's reserves, Satan pleads for "shelter" and shade, the very resources he has helped to eradicate from much of the land. The storm, Satan's last stunt and attempt to deter Christ from his inheritance before carrying him up to the temple's spire, pushes this irony farther—revealing the hypocrisy and self-defeating logic of the unbridled consumption and environmental corruption that he promotes on earth.

A subtler temptation precedes this physical ordeal, however. Before departing the "specular mount" with its piercing views of civilization, Satan steers the conversation to ancient Greece, the birthplace of Western thought and democracy whose history nonetheless is bound up in violent empire building (4.236). Describing Athens, where Satan claims Christ may gain fame and wisdom by cultivating knowledge, he attempts to situate the famed metropolis in an innocent climate, calling it a city "[b]uilt nobly, pure the air, and light the soil" (4.236–84, 239). Despite such claims, the ground and air of this ancient city are clearly not exempt from satanic influence.[97] As the ensuing dialogue between Satan and Christ shows, these environs give rise to a flawed—albeit storied—cultural heritage. Laying out a curriculum of sorts, Satan gives a history of several figures of classical philosophy, poetry, and oratory. Yet, in his account, he inadvertently reveals how their teachings can be made to serve devious ends. For example, in recommending the philosophy of Aristotle, "who bred / Great Alexander to subdue the world," Satan glorifies the formation of a disciple whose relentless military campaigns brought wide swathes of Europe, Africa, and Asia into the Hellenistic sphere (4.252). In Satan's boast that in Greece "thou shalt hear and learn the secret power / Of harmony" there is a faint suggestion, given his predilection for acoustical enchantment, that this tool might be used for compulsion (4.254–55). Such allusions to the darker applications of classical learning belie Satan's stated objective, which is to elevate Christ's mind, not whet his appetite for glory. Satan's concluding suggestion that wisdom in kingship is more complete when it is "with empire joined" confirms that the lessons he intends Christ to draw from classical learning are not pacific (4.284). In pretending to offer Christ a mild, contemplative path to fulfillment while promoting the conflicting goal of empire building, Satan employs the "subtle shifts" and sophisms for which Christ condemns Greco-Roman rhetors (4.300–321, 308). The ancestor of powers such as Rome and Parthia, Greece supplies not only the culture of empire that Satan explicitly promotes in previous temptations, but also the art of eloquence to paint over the cruel and environmentally damaging business of conquest. Christ roundly rejects these Athenian refinements, accusing Satan, in his wonted acoustical and meteorological capacities, of acting through the charming music and poetry of Greece ("thy loudest sing / The vices of their deities") and its famed philosophical schools:

> Who therefore seeks in these
> True wisdom finds her not, or by delusion
> Far worse, her false resemblance only meets,
> An *empty cloud*.
>
> (4.318–21; emphasis added)

Since Christ refuses to accept his feigned offer of learning with its sequel consequences of intellectual confusion and cultural decay, Satan abandons persuasion and allurement in his next tactical move, opting instead for an approach that brings the physical force of the atmosphere to bear on his opponent.

QUELLING THE SATANIC EARTH CYCLE IN THE AIR

Taking Christ back to the wilderness, a place of trial and purification, Satan exercises the most awesome manifestation of his elemental power by calling up a storm in an attempt to shake the Son's determination. David Quint likens the scene to the banquet episode, arguing that both exemplify the brief epic's "removal of the demonic from nature."[98] Not an ordinary exorcism—a rooting out or a chasing away of devils—the expulsion in Quint's account is intellectual, a casting out of human belief in the "sham" idea that the devils' "possess and control the elements."[99] In the alimentary temptation the devils are merely "masquerading" as nature spirits, since, he argues, they are not themselves "animating parts" of nature and Satan's dominion over the air is a "fiction."[100] He questions whether the devil is really responsible for the main meteorological event of book 4 and concludes that the "storm may just be a storm."[101] Despite the fact that Satan possesses the atmosphere illegitimately, his power to manipulate the air and wield the elements for his purposes is nevertheless utterly real. We do not have to take Satan's word for it. When he lifts Jesus up through the atmosphere to the mountaintop, the narrator states that "such power was given him then," and explains, again, when Satan flies to the wilderness with Christ in tow, that "still he knew his power / Not yet expired" (3.251, 4.394–95). The words "then," "still," and "not yet" in these verses suggest the finitude of Satan's atmospheric power but forestall its cessation until a future date. In conformity with the ending of the *Nativity Ode,* Christ's triumph at the conclusion of the brief epic does not immediately save humanity from the power of the air. The angelic choir predicts that the "infernal serpent, shalt not long / Rule in the clouds," implying that his usurped "rule" over the air is still divinely permitted—the demonic has not *yet* been removed from the world (4.618–19). Milton, moreover, describes Satan's atmospheric stunts in concrete terms. As if to say, "I'm not making this up," he explicitly distinguishes satanic flight from that of Romance, as when the devil "caught [Jesus] up, and without wing / Of hippogrif bore through the air sublime" (4.541–42).[102] As we have seen in *Paradise Regained* and throughout Milton's poetry, moving, traversing, plundering, disposing, and combining with the elements, particularly the air, are all within Satan's power.

A common thread throughout the stones-to-bread, banquet, kingdoms, and diabolical storm temptations is their ramifications for the environment. In the first two encounters, we saw that Satan's apparent interest in dispelling scarcity and hunger are masked attempts to implicate Christ in his exploitation of the elements. First, he encourages him to feed himself and others by tapping into the power of stones—an infernal fuel and potential source of pollution and meteorological perturbation. In the sequel episode, he pressures Christ to accept gifts from "Nature ashamed"—food harvested on severe and demeaning terms. The temptation of the kingdoms urges Christ to accept, as a condition of empire building and the accumulation of power and wealth, the price of destroying the lands of conquered people. Although some readers have seen the theme of prosperity as tying together these earlier temptations and, thus, setting them apart from the storm scene, every temptation uniformly focuses on human and environmental loss. Satan's raising a storm to frighten Christ differs from the foregoing temptations, mainly, in that it openly embodies rather than surreptitiously conceals the environmental corruption on which all of his offers are based.[103]

The gale comes at night after Jesus is deposited by Satan in the wilderness, and

> Hungry and cold betook him to his rest,
> Wherever, under some concourse of shades
> Whose branching arms thick intertwined might shield
> From dews and damps of night his sheltered head.
>
> (4.403–6)

Reversing Satan's empty wish that the Son of God would be his "shelter," the passage suggests that Christ's head *was* "sheltered." But the word "[w]herever," used elliptically to emphasize that this episode has no scriptural precedent, also indicates that Christ might have rested anywhere and still could not have avoided the devil's atmospheric blitz. In the next line, the tree branches that the text imagines "might shield" Jesus, providing a physical version of the spiritual shield in Ephesians 6:16, are dismissed as ineffectual: "But sheltered slept in vain, for at his head / The tempter watched" (4.407–8). Recalling how in

Paradise Lost Satan evaded the angelic guard and found his way to Eve's sleeping head, the lines imply that no place or shelter on earth, not Paradise nor the sleeping mind, is or ever was safe from the prince of the air. Regardless of whether the Romans' land is "screened" by hills, Satan's "cold Septentrion blasts," like his airy microscope, can and will overcome the "mountains interposed" (4.39).

The tempter's most explicit meteorological attack on Christ is both physically and mentally threatening. Since Satan holds sway over the elements, we find several of them joining in the mayhem. From the riven sky pour down "[f]ierce rain with lightning mixed, water with fire / In ruin reconciled" (4.412–13). Turbulent air adds to the fray. Recalling the winds' coordinated fury in *Paradise Lost* after astro-geological changes occur in the fallen world, the winds of this satanic storm (a species of that original spate of disastrous weather)

> rushed abroad
> From the four hinges of the world, and fell
> On the vexed wilderness.
>
> (4.414–16)

Like the Parthian army ravaging the hinterlands or Roman gluttons peeling their provinces for riches, the diabolical winds tear through the trees enshrouding Christ. Even the "tallest" and "sturdiest" trunks "[b]owed their stiff necks, loaden with stormy blasts, / Or torn up sheer" (4.416–19). Afterwards, Satan's ironic characterization of the storm as a "sneeze" cruelly mocks its disastrous effects on the ground (4.458). We know that storm raising was believed to be one of Satan's primary capabilities in the fallen world, and in some early modern artistic renderings of the practice the devils themselves appear within the squall, aiding and guiding the winds.[104] Milton follows this tradition by depicting Satan's meteorological agents in their true shapes, riding the storm and attempting to terrify Christ: "Infernal ghosts, and hellish furies, round / Environed thee, some howled, some yelled, some shrieked" (4.422–23). The suggestion that Jesus was "[e]nvironed" by the shrieking storm-devils builds on Milton's spatial and aural representation of affliction (Sin's and his own) as the experience of being submersed in a noxious climate and orbited by terrifying and dissonant sounds.[105] Since Jesus was evidently the eye of the wilderness storm, Satan argues that he should view it as an omen specific to him.[106] But the event is not really an anomaly. As the other temptations demonstrate, the whole world is environed by such furies, which scar the land and make a corrupt and deceitful din through the oracles. No place on earth is exempt from environmental demonism.

Jesus's antidote to this most grave tempest is a sense of steadiness anchored by conscience and reason. The victory of these staid principles over satanic subterfuge in *Paradise Regained* would appear to support Quint's argument that the brief epic rejects magic and embraces a disenchanted scientific worldview.[107] But rather than suggesting Milton's late conversion to Enlightenment ideals, Christ's invulnerability recalls the semi-talismanic effect of chastity in the earlier work *Comus* where virtue prevails over spiritual evil, not by denying the reality of demonic magic, but by drawing on a superior power. Amidst the swirl of wind, ghosts, and lightning, Christ "[s]at'st unappalled" because he possesses "sinless peace," strength-giving knowledge that he has not succumbed to Satan and disappointed the Father (4.425). Instead of condemning the entire premise of meteorological providentialism, which Milton credited along with many contemporaries, Christ's denunciation of Satan's predictions centers on the fiend's nefarious motivations (4.489–98). Without denying that Nature or even diabolical storms may contain portents (indeed, some of what Satan foretells at 4.478–80 is actually true), Christ rejects the devil's interpretation of the tempest because he knows that the storm was not commanded by God. "Thou art discerned," Christ concludes, explaining that Satan "storm'st refused" because he was angered by Christ's refusals and sought to "terrify / Me to thy will" (4.496–97).

If Milton allows Satan's words to portend anything with accuracy, then Satan's claim after the storm to have "heard the rack / As earth and sky would mingle" prefigures his final challenge to Christ on the pinnacle and the accompanying simile on the wrestlers (4.452–53). Earth and sky represent the usurped territories that Satan aims to defend and two crucial locations in the fight between Hercules and Antaeus. That air and earth "mingle" in Satan's words foreshadows the meteorological symbolism of this epic wrestling match wherein Hercules gains the upper hand in the air and Antaeus is aided by Earth, his mother. Alleging that he "heard" the catastrophic storm, a characteristically Miltonic aural rendering of a meteorological event, Satan seems to have a premonition of the sound by which he will perish. As foretold by the singing angels at the end of the brief epic, Christ will chase Satan "with the terror of his voice / From thy demoniac holds" (4.627–28). But before he wages this terrifying aural warfare, Christ stakes out a quieter, if defiant, stance on the spire of the Jerusalem temple. The prince of the air must be refused in his own conquered realm before the Savior begins the aural work of cleansing the air of demons.

Climbing through the "air sublime" and arriving at the temple, Satan sets Jesus down on the highest point of the spired roof and challenges him to "[t]here stand, if thou wilt stand," adding scornfully, "to stand upright / Will ask thee skill" (4.551–52). Assuming that to do so would be impossible, Satan continues to taunt,

> if not to stand,
> Cast thyself down; safely if Son of God:
> For it is written, he will give command
> Concerning thee to his angels, and in their hands
> They shall uplift thee, lest at any time
> Thou chance to dash thy foot against a stone.
> (4.554–59)

What happens next has consumed an enormous amount of scholarly energy. Jesus refutes Satan and unceremoniously stands. The two most debated aspects of this temptation—whether Jesus knows that he is "more . . . than man" and whether his divinity plays a role in helping him balance on the pinnacle—are not the sole sources of its relevance within the brief epic and the poetry overall (4.538). The verses that should help to decide the critical debate, moreover, may be ultimately impossible to disambiguate. Christ's defiant answer to Satan's command to cast himself down, "Also it is written, / Tempt not the Lord thy God," may be construed as an announcement of his deity; but it is equally possible that he means merely to quote scripture (Deuteronomy 6:16) to condemn Satan's blasphemy and defend his own choice as a human being (4.560–61). Fortunately, the text is much more forthcoming about other salient questions regarding, for example, the pattern of heroism exemplified by Christ's actions and the fate of environmental corruption in the world.

Satan's allusion to Psalm 91:11–12, which suggests that the angels are charged to protect Christ even from dashing his foot against a stone, recalls several other mineral-based comparisons in the poem. Despite employing figurative and literal stones in his temptations, "[r]ocks whereon greatest men have oftest wrecked" and grist to make Christ's bread, Satan's attempts are shallow excavations into the deep bedrock of Christ's faith (2.228). Far from tripping over these futile etchings, Christ portrays himself as "a stone that shall to pieces dash / All monarchies" (4.149–50). The comparison of Christ repulsing Satan's arguments to "a solid rock" that shatters the relentless onslaught of "surging waves," also undermines Satan's appropriation of lithic power (4.18). Yet in Satan's final speech of the brief epic, he persists in identifying himself with the "stone" on which Jesus could dash his foot or perish should he fall. The wave analogy offers further insight to why Satan returns so often to stones. In that figure Satan embodies the hectoring waves that, after they break and splinter against the immovable rock, end in "froth or bubbles" (4.20). This process is a cycle that recurs repeatedly in the brief epic. The waves depict the elemental strife and destruction into which Satan strives fruitlessly to draw Christ, only to return to the air, his first possession, deflated and reduced to a mere bubble. Throughout *Paradise Regained* Satan sustains this back-and-forth movement, oscillating between the "rocks" of the earth and the middle air where he consults with his peers, flies, and makes weather. His motion embodies the artificial and natural processes of excavation and exhalation by which subterranean substances pass into the air, cause corruption, and fall again to earth. The text concludes with a single, prophetic, and abruptly truncated round of this cycle.

The Jerusalem temple is an enormous stone, "appearing like a mount / Of Alabaster," whose top reaches up to the "high places" mentioned in Ephesians 6:12 where spiritual wickedness dwells (4.547–48). Satan stresses the height of the edifice when he sets Jesus down: "highest placed, highest is best" (4.553). He brings Christ to this great pile, which is conspicuously crowned with gold, because of his misconception of power as deriving from the elements. If the Son refuses to reveal his divinity *there*—raised above the sparkling stone structure where he would seem most powerful and justified by the place, "thy Father's house"—then, Satan is willing to bet, Christ does not possess divinity (4.552). However, Milton's victor is not a ruler of this world, but a hero in the Pauline-Augustinian mode. Satan fails to see that "high places" such as the Temple spire cut deep into his airy realm where the uncompromising, spiritually armored Christian can inflict the severest blow. In the same way that Augustine's heroes earn their stripes by refusing to appease the powers of the air and conquering them with God's valor, Christ repeatedly denounces the prince of the air's false economy of gifts and tribute, and finally vanquishes him in the heart of the corrupted atmosphere. As Forsyth observes, Ephesians's injunction to arm oneself against principalities and powers "'And having done all, to stand' is precisely what Milton makes Christ do."[108] Christ admonishes his foe and does not budge from the pinnacle. "[H]e . . . stood," and Satan "smitten with amazement fell" (4.562).

The subsequent long and rich simile on Hercules's battle with the giant Antaeus portrays the dynamic between Christ and Satan and its final resolution in stark elemental and cosmographic terms.[109] The analogy occupies six lines to the Sphinx simile's four, suggesting its complexity and significance as a mirroring device for the pinnacle scene and the work as whole. It lingers over the details of the wrestlers and the multiple rounds of their fight, concluding with an additional three lines that explicitly tie the myth back to Satan's attacks and his ultimate demise:

> As when Earth's son Antaeus (to compare
> Small things with greatest) in Irassa strove
> With Jove's Alcides, and oft foiled still rose,
> Receiving from his mother Earth new strength,
> Fresh from his fall, and fiercer grappled joined,
> Throttled at length in the air, expired and fell;
> So after many a foil the tempter proud,
> Renewing fresh assaults, amidst his pride
> Fell whence he stood to see his victor fall.
>
> (4.563–71)

Signaling that Hercules is the son of Jove, ruler of the sky, and Antaeus is "Earth's son," the child of Gaia and Neptune, the passage emphasizes the characters' parentage because the elemental and spatial implications of the comparison are paramount. Scholars have noticed this emphasis of the simile before. Northrop Frye draws a location-based parallel between Christ's last stand and the atmospheric setting of the myth: Satan "is defeated in his own headquarters, the lower heaven or element of air," and the simile tells a comparable story of how Hercules "overcame the monstrous son of earth in the air."[110] Kathleen M. Swaim complicates this parallel, however, observing that "[i]n *Paradise Regained* Christ's victory goes beyond Hercules' by defeating the enemy in the domain of his strength rather than by isolating him from his elemental support."[111] Though Swaim dismisses, I think, unduly, the relevance of Earth's "elemental support" to the pinnacle temptation, she helpfully underscores an apparent discrepancy between the simile and the temptation scene. There are *two* key nodes in the Hercules-Antaeus myth, the air and the earth, whereas Christ's victory over Satan high up in the sky seems to concern the *single* location of the air. We must try to account for why Milton's simile has this double focus and ask why it aligns Satan's counterpart with earth, and not air.

Indeed, Antaeus's connection to the earth is the primary focus of the passage until Hercules delivers the deathblow. Antaeus "still rose" though he was often knocked to the ground, "[r]eceiving from his mother Earth new strength," and after each resurrection, he joined the combat with more ferocity. Lucan's version of the myth offers insight into what it means for Antaeus to acquire strength from his mother.[112] As the legend is told in *The Civil War* (4.593–660), Antaeus is one of Earth's sons, a hulking giant who lives in a cave in Libya where he sleeps on the bare earth to recover his strength, which he spends killing lions and murdering people. Hercules wrestles him, and the first time Antaeus falls to the ground "the dry earth eagerly drank his sweat, his veins were replenished with warm blood, his muscles swelled out, his whole frame grew tough, and he loosened the grip of Hercules with fresh strength."[113] His mother infuses "warm blood" through his veins. We are reminded of the terrestrial "veins" that supply "liquid fire" to the devils' gold mining operation in *Paradise Lost* whose issue "[a]non out of the earth a fabric huge / Rose like an exhalation." Antaeus's ascent, like the raising of hell's temple, is meteoric in every sense of the word. In *Paradise Regained,* his strength was "new," he emerged "fresh," and the fight was "fiercer" after he sprang from the ground. A cave dweller who took the precaution of "pour[ing] hot sand over his limbs" before wrestling Hercules, Antaeus knew the hidden virtue of rocks.[114] Ancient and early modern people were well aware of invigorating materials such as sulfur that lie underground. Lucan's contemporary, Pliny the Elder, classifies sulfur as "among the other kinds of earth the one with the most remarkable properties" and "which exercises a great power over a great many other substances."[115] He lists its many medicinal applications, and, arguing that sulfur has an abundance of fire within it, adds that thunderbolts and lightning have a sulfurous smell and luminescence.[116] As an emblem of elementally renewed strength, Antaeus's relationship to Earth thus reflects Satan's efforts in *Paradise Lost* and *Paradise Regained* to bolster and assert his power through activities such as mining, concocting gunpowder, storm raising, and seeking to convert coal into food.

Without being overthrown, in Lucan's tale Antaeus falls down a second time voluntarily, and this leads to the earth's injury: "[a]ll the vital power that resides in the earth poured into his wearied limbs; and Earth suffers in the wrestling-match of her son."[117] His actions accord with the self-serving philosophy Satan voices during the first temptation, that Christ should transform stones into food to "save thyself" instead of abstaining or waiting to be fed by God (1.344). Lucan's description also carries an echo of Pliny, who lamented the avarice and dissolution that gold brought to civilization and the wreckage that delving in the ground for precious metals wrought on the earth.[118] The Antaeus myth thus has a built-in moral that suits Milton's ongoing critique of Satan as an environmental scourge. Ransacking mother earth to harness the unadulterated power within may have a temporary benefit, but in the long run both the parasitic child and the overtaxed parent suffer.

As our own generation knows too well, reckless mining and overconsumption of the earth's fuels has a dire effect on the air. Milton and his contemporaries recognized this danger as well. The traditional and "gunpowder" schools of meteorology acknowledged, and the sooty condition of London's air demonstrated, that exhalation and sulfurous fumes—either escaped from the ground or emitted from sea coals—could produce noxious pollution, storms, apparitions, thunder, and lightning. The arc of Satan's passage in Milton's epic poetry from an infernal dungeon surrounded by metals, fire, and sulfuric rock to the earth's middle atmosphere, moreover, corresponds to the subterraneous origination of coal smoke and other meteorological phenomena. In terms reminiscent of the slow-acting climatological curse that Milton associates with the Fall, Evelyn describes the effect of releasing these buried fuels into the open: "For all *subterrany* Fuell hath a kind of *virulent* or *Arsenical* vapour rising from it; which, as it speedily destroys those who dig it in the *Mines;* so does it by little and little, those who use it *here* above them."[119] In *Paradise Regained,* as I have argued, this climatologically significant pathway from earth to atmosphere and vice versa is well trodden by Satan and remains a central focus of the poetry. With Antaeus repeatedly availing himself of earthly fuel and the analogy culminating in a mid-air struggle, the Hercules-Antaeus simile offers a schematic distillation and resolution of this corrupted meteorological cycle.

Pivoting sharply from depicting Antaeus's rejuvenation by the earth to showing his deprivation of life-giving air, the simile terminates with a single line: "Throttled at length in the air, expired and fell." Milton's readers, who were well versed in mythology, would discern in this line the climax of the fable in which Hercules succeeds in killing Antaeus by picking him up off the ground and cutting him off from the support of his mother Earth. But Milton departs from classical sources such as Ovid, Apollodorus, and Lucan in portraying Antaeus as "[t]hrottled" by Hercules rather than snatched away from the earth, hugged to death and broken, or gripped by the breast.[120] There is precedent for Milton's focus on stricture and expiration in Natalis Comes's *Mythologiae* (1567), and the poet Abraham Fraunce mentions Antaeus's strangulation.[121] Yet the word "[t]hrottled" is an unusual addition to this story (though perhaps not as unusual as Spenser's "scruzd"[122]). Etymologically descended from the noun "throat," Milton's term emphasizes asphyxiation by choking rather than hugging or squeezing.[123] The word's association with organs of breath and speech was quite strong. Early modern anatomical and medical texts often employed "throttle" to describe the throat or larynx, and according to one account of a pickpocketing incident, a group of thieves assails an apprentice and "thratled him so sore by the wind-pipe, that he could make no noise."[124] Connoting an act of silencing, the word "throttled" thus describes a highly appropriate requital of Satan's atmospheric warfare. From the beginning Satan's abuses on earth have been largely acoustical in character. My analysis of *Paradise Lost* demonstrates that Satan manipulates and speaks through biological organs using methods from the musical sphere. Boasting that he "glibbed with lies" the tongues of King Ahab's false prophets, Satan continues to use this acoustical approach in *Paradise Regained* to shape human history (1.374–75). Christ's declaration at 1.454–59 that henceforth Satan will no longer be permitted to speak through the oracles—which says nothing, however, about his ability to whisper elsewhere—anticipates the kind of victory over Satan that the Hercules-Antaeus simile represents, particularly in its stark final line. Since he is effectively absent from line 568 (Antaeus is only mentioned once, a full five lines earlier), the subject is insistently embodied by the verbs "[t]hrottled," "expired," and "fell." The effect is that the line diminishes the subject to a throat or windpipe, the very organ that Satan uses in his temptations. Similarly, the line does not directly describe who or what crushes this organ, sealing it off from the animating power of the air. We are left to ponder whether a single clenched hand or the pressure of some transcendent, inexorable force is behind this act of suffocation.

Comparing Satan's defeat to Antaeus's asphyxiation in mid-air captures the irony of the situation. The prince of the air has corrupted the primary element from which he derives his power. Satan's throat, the source of so much "inspiring venom" and foul air, finally seizes up in the poisonous climate he has cultivated. In discussing the health effects of coal pollution in London, Evelyn similarly portrays the respiratory and vocal organs as simultaneously admitting and exhaling corrupted air. Representing chimneys and smokestacks as orifices, Evelyn impugns manufacturers such as brewers and soap boilers for "the *Columns* and *Clowds* of *Smoake*, which are belched forth from the sooty Throates of those Works."[125] He urges Charles II to remove these polluting "[t]hroates" to outside of the City lest the human throats of London continue to choke on their poisonous discharge. Likening coal smoke to fumes from hell, Evelyn explains that "the inspiration of this infernal vapour, accompanying the *Aer*" flows into the lungs, "violating, in this passage, the *Larynx* and *Epiglottis*, together with those multiform and curious Muscles, the immediate and proper Instruments of the *Voyce*, which becoming rough and drye, can neither be contracted, or dilated for the due modulation of the Voyce."[126] Satan had been brought "to shameful silence" earlier in the brief epic, but was always able to find a retort (4.22). Now, the lying oracles, howling fiends, and noxious storms with which Satan has filled the atmosphere hamper his own voice as he falls silently from the air in humiliation.

A final consideration is whether Christ's act of standing up to Satan in and of itself changed the region of the air. That Satan damaged the climate enough to strangle himself suggests he would fall even without Christ's direct intervention. The similes support this possibility: Satan's vanquisher Hercules/Christ recedes from view, and instead of being killed, the Sphinx/Satan elects to "[c]ast herself headlong" down to its death (4.575). Yet regardless of whether or not Satan's demise is self-inflicted, there remains a question about the climatological effect of Christ's action. Does balancing on the temple spire purify or liberate the air, and if so, how? To rephrase the question in terms of the new meteorology, which influences Milton's portrayal of the weight of the air in *Paradise Lost*, we might ask whether the barometric pressure changes when Christ admonishes Satan to "[t]empt not the Lord thy God." Earlier in the brief epic, Christ anticipates the restoration of a more just equilibrium between good and evil when he warns Satan, echoing the physics of barometry, that "my rising is thy fall" (3.201). In order to earn his exaltation, Christ also acknowledges that humankind's sins' "[f]ull weight must be transferred upon my head" (1.267). Far from defying the laws of gravity as he balances on the pinnacle, thus, Christ's position represents the shouldering of a universal weight. Heavy sins that the Accuser wields over humanity shift their weight to Christ's body. Like the standing column of mercury in the barometer—what the Royal Society calls the "ballance of the air"[127]—Christ is the counterpoise to the whole weight of the atmosphere whose downward thrust represents the collapsing realm of Satan. Invisibly supported by an unknown combination of inner or outer forces, Christ's body appears miraculously steady much like the barometric column of mercury, whose suspension initially baffled the scientific community.

As barometers became more common in England, a nomenclature emerged for recording where the mercury stood at any given time. Henry Oldenburg's letter to Robert Boyle on October 27, 1664, contains the first known instances of the word "station" as a term for the level of the mercury in the tube.[128] A mutual friend of Boyle and Milton, Oldenburg reports that at Gresham College the "station of the Barometer" was recently at an all-time low and that the operator of the barometer was commanded to record in a diary "the stations of that Mercury, together with the state of the winds and Weather."[129] Similar to the tenuous station of the barometer, Jesus maintains an "uneasy station" on the pinnacle until relieved by the "soft" pressure of angel wings which "upbore" him even higher into the air (4.583–84). He stands on his own, not by superimposing divine power onto Nature, but in spite of what previously seemed impossible by Nature's laws: first by eschewing the need for food, now by finding his balance on the thin spire. The same could be said for the barometer's column of mercury, which produces the so-called impossibility of a vacuum revealing a new physics of air.

Christ's success on the temple cuts across the old and new epistemologies of the seventeenth century. Rather than miraculously superseding Nature's laws or calling on his heavenly Father for aid, Jesus is able to stand on the temple's spire by virtue of a physical climatological change that he initiates through spiritual resistance. In quoting Deut. 6:16 and refusing to claim the filial privileges of the Son of God, Jesus powerfully demonstrates his divinity without actually exercising it, which in turn causes Satan to fall and temporarily relinquish his power over the air (4.576). This transfer of power, which throttles or takes the wind out of Satan, also upholds the Son as he balances on the spire. The effect is similar to that of air pressure on the barometer, which elevates and props up the mercury within. The famous pronoun "him" that Satan shares with Christ in the denouement to the climactic pinnacle scene, and which momentarily suggests that the corps of flying angels rescues the fiend rather than Christ,

> So Satan fell and straight a fiery globe
> Of angels on full sail of wing flew nigh,
> Who on their plumy vans received him soft
> From his uneasy station,
>
> (4.581–84)

affirms the continuity between the force that sends Satan plummeting through the air and that which keeps Christ aloft. Jesus's body meets the shifting conditions of the dispossessed atmosphere with perfect equipoise. Transferring its weight onto him, the atmosphere paradoxically buoys him up.

Allusions to instruments are a recurring feature of Milton's depiction of the fallen climate. These emblems of the New Science typically portray the mechanistic movement of air as a defect of the corrupted climate, as when Satan claims to use an "airy microscope" to magnify and manipulate the appearance of a far-off landscape. In the climactic pinnacle scene, however, the implicit allusion to the "baroscope" the seventeenth century's *other* famous instrument of air (which incidentally appears in Robert Hooke's book on microscopy), signals climate improvement.[130] Unlike Satan's invasive optical instrument, the baroscope, or barometer, is *acted on* by the atmosphere and its response constitutes a sign. As the Royal Society was then learning through a variety of reports, the rising of the mercury in the barometer signals the prospect of fair weather.[131] As with an ascending barometric column, Christ's body signifies climate amelioration because it bears the whole atmosphere's impending weight. Through

>   strong sufferance:
> His weakness shall o'ercome Satanic strength
> And all the world, and mass of sinful flesh.
>   (1.160–62)

Marveling that the barometer could gauge the weather so precisely, John Beale declared, in a letter published in 1666, that he could find "nothing so neerly indicative of the change of weather, as this *Ballance*," ranking it above other "Prognosticks" such as the *"Hygroscope,"* a device for measuring the moisture or aridity of air, and the "sweatings of Marble."[132] Readers will recall that the latter weather sign appears in the *Nativity Ode*'s comparison of the oracles' expulsion from their shrines to the sweating of marble, suggesting an ominous correspondence between metaphysical events and natural processes such as condensation and sweat. In *Paradise Lost* and *Paradise Regained*, Milton reaffirms the notion that spiritual activity lends significance to natural phenomena, but his imagery had evolved significantly since the early seventeenth century to reflect intervening changes in pneumatics and meteorology. Christ resembles Beale's modern *"Ballance"* as he stands atop the temple, embodying, in his steadfast relation to the air, an unprecedented and infallible prognostic of climate renewal. Without fully embracing the New Science or declaring the fallacy of demonism, the ending of *Paradise Regained* echoes the hopes of some seventeenth-century reformers and experimentalists in suggesting that humanity, through the grace and example of its most ideal representative, possesses the spiritual resources to improve the air. Though he towers above the earth, Jesus stays grounded on his Father's temple in stark juxtaposition to portentous meteors often associated with Satan such as comets or lightning. His aim is not to terrorize sinners, but to inspire their faith. The barometer is at once mirrored and obviated by Christ's body, which provides not only an auspicious sign of future conditions, but also the spiritual means to release humanity from relentless climate change. Sin was the undoing of the air, and thus moral heroism—not mere science—must rectify the climate.

As we saw in both epic poems, the Archfiend's environmental corruption ranges far and wide, on land and in the sky, blighting himself along with the source of his power. Though Christ is also subject to environmental corruption and change, wandering everywhere, in "whatever place, / Habit, or state, or motion," he is always immutable: "still expressing / The Son of God" (4.601–2). Though at the end of the brief epic Satan is not actually banished from the atmosphere, Christ's symbolic triumph over the prince of the air clears a path to its restoration. Standing unshaken—not merely once, but over and over, in all places and, spiritually, within the human heart—squelches the storm clouds festering overhead by cutting them off from their immoral human-provided fuels. Yet the fruits of this labor of standing are still out of reach. In recovering Paradise, Christ renounces the delusory hope of shielding oneself from environmental corruption, and instead offers a forestalled refuge to those who work to frustrate this abuse. The strongest rebuke of the wickedness in high places is the self-denying act that improves the air for all and permits Nature to keep "what she asks."

Epilogue

The complex interrelated dynamics of the early modern climate—its acoustic and meteorological phenomena, physical corruption, and spiritual agency—are a primary focus of nearly all of Milton's major works from the youthful *Nativity Ode* to *Paradise Regained*, the culmination of his epic poetry. Evincing much more than a poet's propensity for Nature imagery, recurring climatic descriptions embody a fundamental ground-theme of Milton's poetry. Imbued with theological, intellectual, and cultural meaning, the conditions of climate—from air quality to weather and atmospheric sound—help to define the moral parameters of each work and enable the exploration of the Fall from multiple timeframes and points of view. Demonic agents instrumentalize the air to test and afflict protagonists; divine harmony seizes the atmosphere to bring a spiritual salve to the corrupted world; and signs of climate devastation manifest the everlasting penalty of original sin.

Much like the character of weather, Milton's representation of climate is varied and mutable. Conceived during a century of environmental and scientific transformation when change was literally in the air, Milton's portrayal of climate could hardly remain static. The remaking of climate-related sciences, such as meteorology and music, and the palpable deterioration of the atmosphere in coal-burning England, among other developments, presented Milton's contemporaries with new materialist paradigms for understanding the air. Milton's poetry, as we have seen, is alive to these shifting views and yet simultaneously faithful to some of the eldest truisms about the air. As new sciences such as acoustics and pneumatics shook up the received view of the atmosphere, Milton incorporated their ideas and diction into his poetry even while consistently evoking the Christianized Aristotelian worldview with its compelling, poetic conception of the meteorological exhalation.

Though it has not been the chief aim of this study to decipher Milton's leanings toward the New Science, the poetry's repeated evocations of emerging, and even controversial theories of the air demand additional consideration, especially as to the role that scientific experiment is perceived to play in the corruption—and potential amelioration—of the climate. I have said much already of the specific uses of scientific language and imagery in the poems. For example, this book argues that, in *Comus*, insights from the burgeoning field of acoustics help to expose the limits of human reason and aural perception to repel the dangers posed by unhallowed air. In *Paradise Lost*, allusions to the contemporary discovery of air pressure serve to illustrate the fallenness of the corrupted climate, and depictions of meteorologically sensitive acoustic machines suggest that mechanical artifice facilitates the air's subjection to satanic influence. These instances of scientific allusion express a range of sentiments toward the mechanist, experimentalist turn in the seventeenth-century understanding of atmosphere. On one hand, the use of such allusions in poetry that seeks to render the natural world truthfully, even prophetically, would seem to acknowledge the validity of these new scientific claims. On the other hand, Milton's references to discoveries concerning the atmosphere, in many cases, underscore the deflating effect of this new knowledge. If these insights offer enlightenment, then it is not the sort that dispels humanity's guilt for evils manifested by Nature. His representation of the era's unprecedented investigation of the air confronts readers with a stark picture—all the more haunting for its scientific realism—of the damage sin wreaks on the climate and humankind's ability to thrive in it.

Although Milton's poetry respects the scientific basis of new models of atmosphere, it does not portray these approaches as wholly benign or absolute. Mechanism and atomism may indeed be attributes of air, as Milton's contemporaries increasingly acknowledged, but the compatibility of these physical properties with devilish intervention (as we saw, for example, in Satan's use of instrumental techniques to corrupt Eve in *Paradise Lost*) throws the moral foundations of the climate into doubt. Air that is understood to be chiefly corporeal and manipulatable—instead of diaphanous and spiritous—is potentially more prone to engineering, weaponization, and sleight of hand. From Milton's perspective, moreover, the experimentalists' ability to measure the air's temperature, motions, sound-carrying properties, and weight—thanks to innovative instrumentation and a new philosophy of science—diminishes awareness of the animate and spiritual components of air, both of which are vital to his conception of climate. In *Comus*, Milton leaves magic undefeated at the end of the masque, partly because, in his view, it remains a force for spiritual interference in the air that scientific knowledge unwisely discounts and cannot remedy.

At a moment in history when loss—ice melt, coastline erosion, species extinction, among other casualties of climate change—is endemic, reading Milton's poetry feels like finding a centuries-old map that shows the route, with its many wrong turns and shortcuts, by which we arrived at this perishing point. It is less clear whether this remarkable map shows a way out. Does Milton believe the fallen climate might be improved, if not wholly repaired, by human action, or must its cure wait until the Second Coming? Inasmuch as original sin is a permanent curse, always present to be wrestled with but never thrown off completely, the conjoined phenomenon of climate change is impossible for humans to undo on their own. Yet the poetry acknowledges that some progress can occur on earth toward the reclamation of the air. Much of this headway is made by Christ. For example, the silencing of the oracles in the *Nativity Ode* represents a key turning point in the acoustical composition of the air, disambiguating the sound of holy revelation from that of false predictions. The birth of Christianity, the ode suggests, severely hampers—but does not destroy—Satan's ability to fill the atmosphere with misleading utterance. Christ's triumphal victory over temptation in *Paradise Regained* represents another blow to Satan's atmospheric hegemony, but the text forestalls his final extraction from the air until a future unknowable date. While these scripturally inspired events loosen Satan's grip on the climate and promise an eventual endpoint to his reign as prince of the air, ordinary mortals seem to play little if any role in bringing them about. Perhaps it is necessary for Milton to avoid depicting the Son of God in his capacity as Savior as being assisted by human means. Yet the poetry still invites us to draw conclusions from Christ's opposition to, for example, the ecologically devastating excesses of empire as to the conduct and attitudes societies should adopt to limit climate corruption. In addition, the perpetrators and the victims of climate crimes in Milton's poetry—Adam and Eve, and the Lady of *Comus*, respectively—exemplify in their contrition, and in the Lady's efforts to preserve her purity, the small acts of restitution and resistance that prevent the climate from further denigrating the human moral condition and vice versa.

This book has discussed myriad types of climate change. Magic, elemental intermixture, comets, demonic agency, planetary recalibrations, and divine harmony are among the many forces and processes that introduce change into the atmosphere and natural environment in Milton's poetry. Yet these means of climate intervention belong chiefly to Providential, cosmic, or supernatural figures: angels, devils, sorcerers, God, and Earth. Though their sin wrought the first indelible climate change, humans possess few tools to alter its degraded condition. Faith, reason, and sound are the central means by which they attempt to renew or reconstitute the air. This book brings to light a bedrock assumption of Milton's poetry that vocal expression, including poetry and song, is not merely capable of describing the atmosphere, but also is *part* of the atmosphere and thus a force for changing it. Common or false speech lacks this power, but truly inspired discourse, the sound of "undiscording voice," pierces and animates the environment, and, as the Lady avers in *Comus,* threatens to quash whatever evil resides within. Although the notion that words can alter the atmosphere may seem improbable to us, this logic sat well with Milton's contemporaries. Early modern attitudes toward the ecological ramifications of vocal and musical expression deserve greater explication especially as they pertain to literature on the environment. Might a new field of ecocriticism centered on the environmental immediacy of sound and voice spring from such research, uncovering major implications for Nature-based metaphor and other forms of figuration in Renaissance literature?

In Milton's poetry, when speech is organized into harmonious verse, it can be particularly efficacious against climate corruption. But even the Lady, a gifted singer and passionate defender of truth, declines to utter the solemn doctrine that would disarm her tormentor and destroy the threatening climate of the enchanted wood. Counteracting atmospheric deterioration with one's own self-made gusts of "sacred vehemence," as we have seen, is physically and spiritually depleting, potentially even to the point of fatality. It may also bring persecution down on the speaker, causing their words to be drowned out with sounds of savagery as Milton ruminates in an autobiographical section of *Paradise Lost.* Keeping one's soul uncontaminated while striving to combat the dissonance of the air entails personal sacrifice, particularly in terms of deprivation and humility. Implicit in Milton's idea of the epic poet described in the elegiac verse and elsewhere, are instructions for the ecological use of speech. Instead of merely signaling his spiritual worthiness, the poet's obligation to drink only clear water, spurn worldly riches, and lead a secluded abstemious life, reflects a wider conviction that inspired speech must be physically conserved and kept from circulating in the service of waste and corruption.

Indeed, it is ultimately sacrifice—that of the Son of God—that will finally enable the expulsion of Satan's empire from the middle air. In the poetry, this future terminal act of purgation is foreshadowed by Christ's triumph on the temple spire—his final repudiation of Satan's inducements to earthly power and their devastating environmental consequences. I have argued that the choreography of this fateful episode of *Paradise Regained,* in which Christ stands upright as Satan plummets downward, mirrors the operation of the barometer, an iconic invention of the New Science and the herald of a dawning era of modern meteorology. The effect of the comparison is not to equate Christ's stance with a leap forward in scientific knowledge or human progress. Perceptions of the climate were indeed primed to change during Milton's lifetime due to scientific discoveries like that of the barometer, but Christ's triumph over temptation, with its echoes of the Resurrection, offers something far more revolutionary: a pathway to eternal life. Prevailing for the moment over Satan on the temple spire, Christ embodies a new covenant with humanity that promises deliverance from sin in return for unqualified faith. Even as it exemplifies an era captivated by scientific demonstration, the barometer, too, demands the viewer's faith. To accept its measurements, he or she must come to believe that air pressure—the invisible, imperceptible weight of the atmosphere—not another local cause, is the force that pushes the mercury up. In the ambiguity of its functioning, then, and its promise of a new order by which humanity may coexist with the climate, the barometer is a sign akin to Christ's mysterious feat atop the steeple, which alleviates human suffering by prefiguring an eventual end to Satan's reign as prince of the air.

While Milton could see that the tools of the New Science would fundamentally change humanity's relationship to the climate—in the same way that Christ's blow to Satan alters the endgame for climate-embattled Earth and humankind—his implicit allusion to the barometer also expresses the sanguine view that such innovation need not extinguish the memory of original sin in the natural world. Scientific progress can either anchor us to, or untether us from, ancient knowledge of humanity's corruption of the climate. Though in much of the poetry Milton is somewhat leery of the New Science for its moral and philosophical blind spots, he expresses a more favorable view of the future of meteorology in his last epic poem. In portraying Christ's body as anticipating the action of the barometer, the final episode of *Paradise Regained* posits a harmony between the moral dynamics of climate and scientifically measured conditions of the air. Put another way, the epic refuses to decouple the spiritual dimension of the air from scientific representations of the atmosphere. Evermore precise weather prognostics may inform, prepare, or warn, but never release humanity from its moral accountability to the air. Assuming an indissoluble correspondence between divine justice born out through the weather and the new meteorology's scientific determinations about the air, Milton's poetry assures us that the climate is morally comprehensible to all.

Notes

INTRODUCTION

1. Fowler, ed., *Paradise Lost,* 9.1002–4. Henceforth, all quotations of the epic are from this edition and cited in the text by book and line number.

2. London used more mineral fuel than any other city during the seventeenth century. For a comparison of fuel consumption worldwide, see Cavert, *The Smoke of London,* 29–30.

3. For example, Parker (*Global Crisis*) demonstrates that the Little Ice Age, which entered its longest and most severe episode during the seventeenth century, catalyzed an era of unprecedented political and social upheaval across Europe and Asia known as the "general crisis"; for the effect of the climatic downturn on quality of life, see especially 1–25. Although far less attention has been paid to the moral and artistic implications of climate adversity, Behringer, *A Cultural History,* 85–167, considers the cultural consequences of the Little Ice Age in Europe including food scarcity, mortality, and scapegoating. Fink, "Milton and the Theory of Climatic Influence," 70, addresses Milton's shifting attitude towards the contemporary idea that "cold" northern climates dampen "artistic competence."

4. *De Doctrina Christiana,* 8.1:301, 303. Milton calls the animating breath that God infuses into the first human beings "a certain breeze or divine power wafted out" and equates the word *spirit* in scripture with "the breath of life which we take in" (8.1:303). All quotations of *De Doctrina Christiana* are from *The Complete Works of John Milton* (henceforth abbreviated in the text as "*CW*") and hereafter cited parenthetically by volume and page number.

5. Kerrigan, *The Sacred Complex,* 54.

6. *A Political Theology,* 48. Northcott provides a brief history of the formation of the Enlightenment "cosmopolis" and its objectification of Nature on 39–47.

7. Sloterdijk, *Terror from the Air,* 87.

8. Neiman, *Evil in Modern Thought,* 39. Philosophers divide evil into two kinds: *natural* evil, which cannot be attributed to a purposeful agency, and *moral* evil, which can.

9. *We Have Never Been Modern.*

10. Weber, "Science as a Vocation," 30.

11. On Milton's animist materialism in the context of seventeenth-century intellectual and political debates, see Fallon, *Milton among the Philosophers,* and John Rogers, *The Matter of Revolution,* 9, 11, 103–76.

12. Thomas Kuhn coined the term "proto-sciences," to describe the forerunners of the "mature sciences" that do not display clear signs of progress ("Reflections on My Critics," 138). For a helpful overview of histories of early proto-sciences see Smith, "Science on the Move," esp. 349–52.

13. For example, Anderson accepts the view that the "science of weather . . . was modern," explaining that Victorians saw the meteorology of previous generations as an immature science in need of a Newton to bring mathematical laws and certainty to the subject matter (*Predicting the Weather*, 1–7).

14. It is possible only to give a sampling of recent ecocritical studies of Milton. On Milton's participation in contemporary conversations around air science and the horticultural remediation of air pollution in London, see Tigner, *Literature and the Renaissance Garden*, 195–211; for Milton's ecological depiction of prelapsarian life, see McColley's prolific scholarship (e.g., *Milton's Eve; A Gust for Paradise*); a variety of environmentally oriented interpretations appear in Hiltner, ed., *Renaissance Ecology*.

15. See Hiltner, *Milton and Ecology*, and also, Hiltner, *What Else Is Pastoral?*; Marcus, "Ecocriticism and Vitalism in *Paradise Lost*."

16. Throughout the medieval and early modern eras, church doctrine reflected the popular belief that Satan, sometimes in cooperation with witches, could stir up storms to afflict human beings; see White, *A History of the Warfare*, 1:336–38. Illustrating the climatological emphasis of witch lore, historians have shown that during the sixteenth and seventeenth centuries, some of the worst years of the Little Ice Age, climate-related crop failures and food shortages contributed to a sharp rise in witchcraft trials. See, for example, Oster, "Witchcraft, Weather and Economic Growth"; Behringer, "Climatic Change and Witch-Hunting."

17. On climatic zones and climate determinism (the belief that cultures are shaped by their latitudinal position and weather) see Fleming, *Historical Perspectives on Climate Change*, 11–12. For the Greek origins of this idea, see Thommen, *An Environmental History*, 25. On medieval climatic zone theory, see Lawrence-Mathers, *Medieval Meteorology*, 34.

18. E.g., portraying himself as flying above earth in the invocation to book 7, Milton states that he wishes to avoid the fate of Bellerophon who fell from the sky, but "from a lower clime" (18). In contrast to Milton, Bellerophon did not reach heaven and fell from the atmosphere.

19. Rudimentary meteorological research was part of this flowering of scientific inquiry, but the objectives and success of the weather sciences prior to 1900 remain disputed. Golinski, *British Weather and the Climate of the Enlightenment* demonstrates that instruments and the institution of daily weather observation during the eighteenth century helped to found a systematic study of the British climate and "normalize" the weather. But in *Reading the Skies: A Cultural History of English Weather*, Jancović argues eighteenth-century naturalists' reportage of anomalous or "unusual weather" was a development of classical meteorology, not a definitive step towards modern scientific practice. For the epistemological, technological, and sociological conditions surrounding Robert Boyle's production of "pneumatic facts," including the scientific conventions he established to demonstrate the pressure of the air, see Shapin and Schaffer, *Leviathan and the Air-Pump*, 22–79.

20. Mersenne, *The Books on Instruments*, 294.

21. Mersenne, 295.

22. Mersenne, 295.

23. For example, in his depiction of the fragrant "air" that surrounds Eve, the "spirits odorous" exhaled by flowers and their "airy" leaves, and the "worlds and worlds" that Satan and Raphael pass by on their way to earth—seemingly balmy planets that Milton compares with "happy isles, / Like those Hesperian gardens famed of old, / Fortunate fields, and groves and flowery vales" (5.268, 481–82, 567–69; 9.425, 459).

24. Trenberth and Guillemot, "The total mass of the atmosphere."

25. Shakespeare, *Macbeth*, 1.6.5–6, 10.

26. *OED Online*, s.v. "air, n.1," accessed May 19, 2020.

27. Mersenne, *The Books on Instruments*, 294–95 (Mersenne, *Harmonie universelle contenant la théorie*, 5.225–26).

28. Shakespeare, *The Tempest*, 3.2.133–40; for sigh-tempests see, for example, Daniel, "Sonnet XLIII"; Donne, "A Valediction: forbidding Mourning."

29. *Lycidas*, pp. 243–56, 252. All references to Milton's poetry, with the exception of *Comus* and the epic poems, are from Carey, ed., *Milton: The Complete Shorter Poems* and henceforth cited parenthetically in the text by line number. For prose translations of the Latin poetry, page numbers instead of line numbers are indicated.

30. On the concept of pneuma in relation to Stoic god and cosmology, see Meyer, "Chain of Causes," 79–80.

31. Long, "Stoic psychology," 561.

32. *Anatomy of Melancholy*, 1.2.2.5, p. 81.

33. Plato is the fountainhead of this tradition (*Republic*, 475, and *Timaeus*, 67–73, 109). Countless writers in the liberal arts tradition adapted the theme and the Neoplatonists added the notion of the soul's attunement to the celestial music; see Hutton, "Some English Poems in Praise of Music," 145–208, esp. 150–56.

34. As Meyer-Baer, *Music of the Spheres*, 220, explains: "[T]he soul *is* air, the *pneuma*, the *ruah* of the Hebrew text of the Bible. This medium—breath, air, *pneuma, ruah*—is readily related to music, especially to the use of wind instruments."

35. Chapter 2 covers this subject in more detail. See also Gouk, "Music, Melancholy, and Medical Spirits," 173–94, esp. 173–79; Gouk, "Some English Theories of Hearing," 95–113, esp. 101.

36. These *species*, invisible miniature copies of an object or sound, "which are fallen from things, and speeches, and multiplied in the very Aire" proceed from a person's sense organs to the fantasy and finally to the soul where they leave an impression. *Three Books of Occult Philosophy*, 14.

37. As Gouk, *Music, Science and Natural Magic*, explains, the seeds of experimental science were latent in occult traditions, which closely studied music and sound.

38. This idea was enshrined in Boethius's discussion of the *musica mundana* whose harmonic principles were regarded as the highest subject of musical science.

39. Gouk's historically and culturally attentive treatment of the subject of music underscores the sheer number and variety of influences that contributed to its revolution and the simultaneous founding of acoustics at the beginning of the seventeenth century. In addition to the experimental philosophers, musicians, artisans, magicians,

mathematicians, instrument makers, machinists, and others were involved in forging a new understanding of sound (*Music, Science and Natural Magic*, 4–5, 10–14). Several other useful historical investigations describe this shift, for example, Cohen, *Quantifying Music*; Gozza, ed., *Number to Sound*; and Hunt, *Origins in Acoustics*.

40. *Second Defence of the People of England*, 121.
41. *Principles of Philosophy*, pt. 4, §198:281.
42. *The Devil and Demonism*, 2.
43. *Anatomy of Melancholy*, 1.2.1.2, pp. 50, 46, 1.2.2.6, p. 84.
44. The phrase derives from the judgment of Genesis 3:17–19.
45. For a chronological account of scientific and public knowledge about the climate crisis, see Weart, *The Discovery of Global Warming*.
46. Climate change measures the human induced fluctuation that occurs *in addition to* natural climate variability over a comparable period of time; see United Nations, "United Nations Framework Convention on Climate Change," §1:7.
47. Luther, *A Critical and Devotional Commentary on Genesis*, 314. Luther gave the lectures later published as *Ennarationes* (Genesis 3:15 to chapter 9) throughout 1536; see Stjerna and Pedersen, "Introduction to Lectures on Genesis 1:26–2:3 and Genesis 2:21–25, 1535," 71–72.
48. The term "peccatogenic" derives from the Latin word *peccatum*, for sin. On the climate outlook of the colonists, see Kupperman, "Climate and Mastery of the Wilderness," 26–29.
49. This book will repeatedly refer to Ephesians 2:2. I modernize the spelling of "aire" in the Authorized King James Version to "air."
50. *OED Online*, 3rd ed. (2004), s.v., "outrage, n., 1.a.," accessed December 20, 2019.
51. *OED Online*, s.v., "outrage, n., 1.b."
52. *OED Online*, s.v., "outrage, n., 2.a."
53. To quote the *OED*, "[f]ierce and overwhelming indignation, anger, etc., experienced in response to some injustice or affront" (s.v., "outrage, n., 5").
54. In 2019, the United Nations Environment Programme (UNEP) released a self-described "bleak" tenth edition of the *Emissions Gap Report*, which presents a gaping difference between "'where we are likely to be and where we need to be'" in terms of limiting global temperature rise to the levels prescribed by the Paris Accord. The report focuses on the outsized impact of G20 countries, which account for around 78 percent of global greenhouse gas emissions, noting that the failure of large economies to enhance their commitments to meeting the Paris targets, and in some cases, failure to honor existing commitments, will limit global efforts to close the emissions gap. See *Emissions Gap Report 2019*, xiii–xvii.
55. Environmental examples of this attritional and incremental form of violence include the disposal of nuclear waste, acidifying oceans, melting ice sheets and glaciers. See Nixon, *Slow Violence*, 2.
56. For the word "impression" in an early modern meteorological context, see for example, Fulke, *A goodly gallerye with a most pleasaunt prospect*.
57. Lenton et al., "Tipping elements in the Earth's climate system," was the first study to establish these tipping points. In a follow-up study, Lenton and six other

researchers argue that the tipping points now have a higher probability of occurring at lower thresholds of global temperature rise. See Lenton et al., "Climate tipping points—too risky to bet against."

1. "INFANT CRIES"

1. See Milton's 1633 letter "To a Friend" (*Complete Prose Works of John Milton*, 1:319–21, 319). All quotations from this edition (henceforth abbreviated in the text as "*CPW*") are hereafter cited parenthetically by volume and page number.

2. In treating the afterlives of the deceased, mourning poems such as *In Obitum Praesulis Eliensis* (*On the Death of the Bishop of Ely*) (1626), *Lycidas* (1637), and *An Epitaph on the Marchioness of Winchester* (1631), also refer to cosmic themes. In reciting numerous classical precedents for the deaths, these poems, however, stay grounded in ancient lore. Milton's Italian and English sonnets and some of the elegies, poems that were not commissioned, privilege human affairs: erotic love, friendship, spring, society. These worldly concerns are central but typically do not overshadow the speaker's desire for spiritual transcendence (see especially "Sonnet III" and "Sonnet VI").

3. Aristotle's criticism of Pythagoras, Milton points out, rests on the assumption that the doctrine of the music of the spheres is literal.

4. *The Life of John Milton*, 1:286.

5. Hollander, *The Untuning of the Sky*, 30.

6. The notion that the soul possesses a migratory "container" may be traced to ancient hermetic texts and Stoic philosophy, and appears in the writings of Porphyry, Proclus, Macrobius, and others. See Klein, "Spirito Peregrino," 62, 65.

7. Klein, 62–72. The spirit was sometimes conceived of as aerial and igneous as well as astral or ethereal (65).

8. Long, *Hellenistic Philosophy*, 147–75.

9. Long, "Stoic psychology," 561. For the permeation of pneuma throughout the universe see Furley, "Cosmology," 440–41.

10. Furley, 440.

11. Grant, *A History of Natural Philosophy*, 42.

12. Aristotle's sketch of the scope of his natural investigations in *Meteorologica* is ordered vertically, beginning with topics relating to the movement of the heavens, transformations of the terrestrial elements, and moving downward to earth dwelling organisms. See *Meteorologica*, 5–7.

13. Richard Holdsworth, a prominent Cambridge don in the early 1600s, taught Aristotle's *Meteorologica* to students in their fourth year, for example. Meteorology, however, was by no means a primary subject of study at Cambridge. The section on meteorology in Keckermann's *Systema* spans less than one hundred pages in the two-volume set, which is approximately three thousand pages long. For Holdsworth's pedagogy and an in-depth discussion of the seventeenth-century physics curriculum at Cambridge see Costello, *The Scholastic Curriculum*, 42, 83–102, 148–49.

14. Aristotle, *Meteorologica*, 29–33, discusses meteoric phenomena known as "goats," torches, and shooting stars.

15. Kelley and Atkins ("Milton's Annotations of Aratus," 1096–99) define two periods that Milton annotated his copy of Aratus. The first occurred after he purchased the book in "1631-early 1638, perhaps ca. 1631–32," before his departure to Italy. The second period occurs after Milton's trip to Italy and most likely when he was tutoring his nephews: "1639–52, probably ca. 1641–42."

16. *Paradise Lost* 10.661–64 echoes the passage Milton noted from Aratus. See Kelley and Atkins, "Milton's Annotations of Aratus," 1103.

17. A classic in this tradition is Ptolemy's book on astrology (*Ptolemy's Tetrabiblos: Or, Quadripartite Tetrabiblos*), which reveals the art of weather forecasting based on the appearance of heavenly bodies.

18. Evans, *The Miltonic Moment*, 15–16.

19. Aristotle explains the origins of exhalations, which he divides into two kinds (hot-dry exhalation and cold-moist vapor) in bk. 1, chap. 4, which demonstrates that hot exhalation is the material cause of shooting stars. It is also said to play a role in forming certain kinds of comets (bk. 1, chap. 7). Vapors from the earth cause cloud, mist, drizzle, and rain (bk. 1, chap. 9). Snow is simply a derivation of cloud (bk. 1, chaps. 10–11).

20. Taub, *Ancient Meteorology*, 74–75.

21. Anaximenes's views about *aēr* are passed down through Aristotle, Aëtius, and Theophrastus (qtd. in Simplicius). See Taub, *Ancient Meteorology*, 75.

22. See Taub's analysis of pneuma in Anaximander's meteorology and also Aëtius's gloss on Anaximenes's analogy between the breath of the body and the wind of the world, qtd. in *Ancient Meteorology*, 74–75.

23. According to Aristotle, air is "made up of these two components, vapor which is moist and cold . . . and smoke which it hot and dry"; *Meteorologica*, 167.

24. At 2.4.360a21–35 and 2.4.360b1–26 Aristotle discusses the seasonal effects produced over time by the predominance of one exhalation in a region. There are many other places where Aristotle attributes meteorological phenomena to a single exhalation. For instance, at 1.4.341b25–35 he discusses the origins of meteors in the upper atmosphere, which occur because the windy or hot exhalation in that area is inflamed by the movement of celestial bodies. See pp. 30–33 and 167–69.

25. At 2.4.360b24–26, for example, Aristotle suggests that dry and moist exhalations can predominate in neighboring regions of the sky much like different parts of the stomach can be variously moist and dry. Another biological analogy occurs at 2.8.366b14–19, where Aristotle argues that earthquakes are like the tremors our bodies experience from wind trapped within them. See pp. 169, 209.

26. For the Aristotelian notion that hot exhalations within the earth mimic the digestive function of the stomach, see Taub, *Ancient Meteorology*, 99–100.

27. *Natural Questions* 2:81.

28. For example, Liberti Froidmondi uses the term "halitus" for exhalations: "Aristoteles, & vniuersi omnes post eum, except vno Paracelso, materiam omnium meteororum in duobus halitibus constituit; in vapore, calido & humido, qui Grecé ἀτμίς, & exhalatione, calidâ & siccâ, quæ ἀναθυμίασις" (Froidmondi 148). Another seventeenth-century textbook writer, Eustachius a Sancto Paulo, substitutes the word "spiratio" for dry exhalation: "Bene scribit Aristoteles 1. Meteor. 13. materiam ventorum esse spira-

tionem siccam & calidam seu exhalationem" (Eustachius, *Physica*, 244). Eustachius's *Summa Philosophiae* was reprinted at least nine times between 1609 and 1628 (Van De Pitte, "Some of Descartes' Debts," 494n1). The word *"exhalatio"* is used ubiquitously in meteorological literature.

29. Aristotle, *Meteorologica*, 77.
30. Aristotle, 70–71.
31. Aristotle, 77.
32. "The efficient, controlling and first cause [of precipitation] is the circle of the sun's revolution. For it is evident that as it approaches or recedes the sun produces dissolution and composition and is thus the cause of generation and destruction," *Meteorologica*, 69.
33. "The cycle of changes reflects the sun's annual movement: for the moisture rises and falls as the sun moves in the ecliptic. One should think of it as a river with a circular course, which rises and falls and is composed of a mixture of water and air. For when the sun is near the stream of vapor rises, when it recedes it falls again," *Meteorologica*, 71.
34. See Seneca's definition of wind (quoted on p. 32) and also Aristotle's explanation of earthquakes, which he attributes to the generation of wind inside the earth; *Meteorologica*, 205.
35. Pliny, *Natural History* 1:247.
36. In *Timaeus*, 143, Plato states that the circular motion of the All compresses the elements into each other, accounting for their mutual transformations. Likewise, Pliny seems to think that the outer spheres of the world exert a pressure on the matter below, but he adds that some matter strives upward.
37. Pliny, *Natural History* 1:177.
38. Pliny, 1:177.
39. Pliny, 1:249.
40. Pliny at first mentions this distinction when discussing the origin of rain and other precipitation. In a subsequent explanation of thunderbolts he states that their fixed cause is the influence of the stars, and when storms arise from the motions of these heavenly bodies they portend future events. Thunder and lightning that arise from exhalations, on the other hand, are a matter of chance and therefore have no prophetic value. See 1:249, 255.
41. Pliny, 1:255.
42. Pliny, 1:255.
43. We know from "At a Solemn Music" that Milton attributed Orphic power to music that melds words with melody, and in *Ad Patrem* (lines 50–55) he implies that words and music are more powerful when mixed than they are alone. Nature's speeches may be inferior because they are unmusical.
44. Hollander, *The Untuning of the Sky*, 26–27.
45. Hollander, 24–25, 28–29.
46. See Finney, *Musical Backgrounds*, 102–25.
47. Finney, 106.
48. "Assuredly, the world's body is living in every part, as is evident from motion and generation"; "When it [music] imitates the celestials, it also wonderfully arouses

our spirit upwards to the celestial influence and the celestial influence downwards to our spirit. Now the very matter of song, indeed is altogether purer and more similar to the heavens than is the matter of medicine. For this too is air, hot or warm, still breathing and somehow living; like an animal, it is composed of certain parts and limbs of its own and not only possesses motion and displays passion but even carries meaning like a mind, so that it can be said to be a kind of airy and rational animal"; *Three Books on Life*, 3.3.255, 3.21.359. See also Finney, *Musical Backgrounds*, 103, 105–8.

49. Furley, "Cosmology," 440, 444.

50. Ficino, along with certain Neoplatonists, and some Hermetic works, saw music as capable of transferring celestial and prophetic influence to man, and even inanimate things, because of its natural sympathy to the cosmic order. See Finney, *Musical Backgrounds*, 105–9.

51. Evans, *The Miltonic Moment*, 33.

52. Finney, *Musical Backgrounds*, 104.

53. See the excerpted letter by Beale in *"Observations continued upon the* Barometer," 164. See also s.v. "Signs of rainy Weather by solid bodies": "stones, especially marble, will sweat against wet Weather, though it be from an outward cause" (Miller, *The Gardeners Dictionary.*)

54. Milton mentions Hermes and Plato directly before the doctrine of the elemental demons. Such spirits are discussed in the Hermetic corpus, Marsilio Ficino's commentary on Plato's *Symposium*, and the Orphic hymns. See Carey, *Milton: The Complete Shorter Poems*, 148n93.

55. Lucretius, *De Rerum Natura*, 529.

56. Milton, *A Mask Presented at Ludlow-Castle*, 110, lines 800, 801–4. All quotations of Milton's masque (*Comus*) are from Milton's 1645 *Poems* and henceforth will be cited in the text by line number.

2. EARLY ACOUSTIC THEORY AND THE AURAL SOUL IN *COMUS*

1. See for example, Marcus, *The Politics of Mirth*, 169–212; Norbrook, "The Reformation of the Masque," 94–110; and McGuire, *Milton's Puritan Masque*. In Marcus's analysis, *Comus* purposely raises and then denies the generic expectations of the court masque in order to distance Egerton's presidency of the Council of Wales from Whitehall and encourage his independence from the ecclesiastical overreach of Archbishop Laud. Norbrook (106) and McGuire (6) view Milton's *Comus* as a "radical" antidote to the dissolution and royalism that the conventional court masque had come to symbolize.

2. Laud's church reforms and the reissuing of *The Book of Sports* (1633) are oft-cited contexts for Milton's critique of Cavaliers in the masque. Breasted, "'Comus' And The Castlehaven Scandal," asserts the relevance of the infamous trial and execution of the Earl of Castlehaven, the Countess of Bridgewater's brother-in-law, and exemplifies a group of studies that relate thematic elements of the work to the family affairs and judicial responsibilities of the masque's main honoree, John Egerton.

3. Brown, "Milton's Ludlow Masque," 28.

4. For example, in the university exercises, *At a Vacation Exercise* and *Prolusion II*, as discussed in chapter 1.

5. See, for instance, the invocation to book 7 of *Paradise Lost* where the narrator metaphorically descends from heaven, confronts earth's barbarous acoustics, and asks Urania to protect his song (30–39).

6. Orgel, "The Case for Comus," 33, argues that the young Egertons' speaking parts are evidence that the Ludlow masque was not performed for a public audience. But the masque's self-consciousness about the act of speaking out suggests that the author and performers were aware of the unusualness of their production for the liberties it gave the young people.

7. Brown, *John Milton's Aristocratic Entertainments*, 78. Brown argues that the decision to give the children full dramatic speaking roles was unusual—it added to the educational effect of the masque and created a sense of realism that "came close to being subversive of the form" (78–103, 78).

8. Ficino produced a Latin edition of the *Corpus Hermeticum* in 1463 (first printed in 1471), and his *Opera Omnia* was published posthumously in 1576. Agrippa's *De occulta philosophia* appeared in 1533 and was translated into English in 1651.

9. For a discussion of the traditional academic curriculum at Cambridge, see Costello, *The Scholastic Curriculum*. Scholasticism was more entrenched at Oxford than Cambridge, though the universities' curricula did not differ much (2–3). Gouk notes that although Bacon condemns Scholastic methodologies, he relies substantially in *Sylva Sylvarum* on Aristotelian explanations of sound (*Music, Science and Natural Magic,* 159–60). Martin argues that Descartes' *Les Météores* continues, rather than breaks with, traditional Aristotelian meteorologies by excluding formal and final causation from its procedures ("Causation in Descartes' *Les Météores,*" 217–36). Galileo's *Discorsi e dimostrazioni matematiche intorno a due nuove scienze* (*Two New Sciences*) (1638) examines problems posed in the Aristotelian *Mechanica* (Garber, "Pre-History of the Mechanical Philosophy," 13).

10. "Before the intellectual status of mathematicians had been raised by Copernicus and other Renaissance *mathematici* there was natural philosophy and there was mathematics and they were essentially separate and distinct.... By the final decades of the seventeenth century ... the notion that there could be mathematical principles of natural philosophy could be taken for granted"; Henry, *The Scientific Revolution*, 28. For a concise history of how the revolution in astronomy, among other factors, led to the mathematization of natural philosophy, see 14–29.

11. On Digges's rendition of Copernicus's cosmology, published in a *Perfit Description of the Caelestiall Orbes* (1576), see Danielson, "Astronomy," 213–14.

12. Garber, "Pre-History of the Mechanical Philosophy," 25–26.

13. Fallon examines the particular brand of Platonism *Comus* displays in *Milton among the Philosophers*, 81–83, but does not mention possible influences such as the occult philosophy or contemporary developments in experimental science. Duran places *Comus* in the scientific climate of the seventeenth century, reading the rational conduct of the Lady in the woods as reflecting shifts in educational practices for noblewomen, ornithology, and the study of sound. Many of the scientific sources Duran deals with postdate the masque, and though Duran draws a brilliant comparison between the

Lady's aural comportment and the tenets of Francis Bacon's acoustical program, no other acoustical theory is mentioned except the promise of distant research in "the second half of the seventeenth century"; see *The Age of Milton and the Scientific Revolution*, 225–49. Edwards focuses primarily on the scientific contexts of *Paradise Lost* in *Milton and the Natural World*, as do Picciotto, *Labors of Innocence* and Marjara, *Contemplation of Created Things*.

14. On the Aristotelian organization of the mixed sciences and their eventual elevation in social, academic, and intellectual spheres, see Gouk, *Music, Science and Natural Magic*, 82–84, and Henry, *The Scientific Revolution*, 18.

15. Rossi, *Francis Bacon*, 7.

16. Gozza, *Number to Sound*, xi.

17. Mancosu, "Acoustics and Optics," 598–602.

18. Vincenzo Galilei developed an experiment to disprove the claim traditionally attributed to Pythagoras that strings tightened by weights that correspond to the musically "perfect" ratios produce consonant pitches. Galileo, Vincenzo's son, conjectured that consonance, following Giovanni Battista Benedetti (1530–1590), is purely a consequence of the regular coincidence of the pulses of two differently pitched, simultaneously vibrating strings. Marin Mersenne advanced this "coincidence" theory of consonance and defined many other attributes of sound, such as its speed, through experimentation. See Mancosu, "Acoustics and Optics," 603, 605–8.

19. For an account of Vincenzo's major experiments see Mancosu, 603–4.

20. Shumaker, *The Occult Sciences*, 108.

21. Agrippa's categories of ceremonial, celestial, and natural magic were well known. The medieval magician Roger Bacon distinguishes between "real" and artificial magic. Gouk discusses some of these categorizations in *Music, Science and Natural Magic*, 71, 85–86.

22. *Of the Vanitie and Uncertaintie of Artes and Sciences*, qtd. in Rossi, *Francis Bacon*, 18–19.

23. *Three Books of Occult Philosophy*, 167–69, qtd. in Gouk, *Music, Science and Natural Magic*, 85.

24. Henry, *The Scientific Revolution*, 55–56.

25. Gouk, *Music, Science and Natural Magic*, 158.

26. For a summary of Gouk's findings regarding the relationship between music and magic in the sixteenth and seventeenth centuries see *Music, Science and Natural Magic*, 110. Chapter 3 of *Music, Science and Natural Magic*, titled "Intellectual Geographies," charts how the "cognitive domains," as Gouk calls them, of music, magic, and experimental philosophy interrelate (66–111).

27. For a summary of Mersenne's acoustical contributions, whose laws first appeared in *Harmonie universelle* (1636), see Gouk, *Music, Science and Natural Magic*, 170–78; for Dee and Fludd see 86–88, 95–101.

28. Fludd's *History of the Macrocosm and Microcosm* (1617–21) includes descriptions of common instruments as well as musical devices from his imagination. Gouk argues that later writers in the acoustical tradition such as Bacon and Athanasius Kircher who were interested in how instruments project sound drew liberally from Fludd; Gouk, *Music, Science and Natural Magic*, 100–101.

29. Gouk, "Music in Francis Bacon's Natural Philosophy," 136.

30. Bacon, *Sylva Sylvarum*, 385.

31. Furthermore, Bacon's suggestion that a mannequin might possibly be made to imitate speech, recalls the legend of the magical speaking head of the famous medieval magus Roger Bacon. See Gouk, *Music, Science and Natural Magic*, 160, 162, 166–68.

32. On Bacon's belief in the existence of spiritus in all matter and his idea that inanimate things have the ability to "perceive," see Walker, "Francis Bacon and *Spiritus*," 122–23.

33. Gouk, "Music in Francis Bacon's Natural Philosophy," 140. For a brief history of the role of the "species" in explanations of sound perception, see Burnett, "Sound and its Perception," 61–62.

34. Gouk, "Some English Theories of Hearing," 99, 102–3.

35. Walker, "Francis Bacon and *Spiritus*," 127.

36. Bacon, *Sylva Sylvarum*, 389.

37. Bacon, 389.

38. For example, in *Of the Dignity and Advancement of Learning*, Bacon denounces a "degenerate" form of natural magic, practiced in his day: that which "flutters about so many books, embracing certain credulous and superstitious traditions and observations concerning sympathies and antipathies, and hidden and specific properties, with experiments for the most part frivolous, and wonderful rather for the skill with which the thing is concealed and masked than for the thing itself; it will not be wrong to say that it is as far differing in truth of nature from such a knowledge as we require, as the story of King Arthur of Britain, or Hugh of Bordeaux, and such like imaginary heroes, differs from Caesar's Commentaries in truth of story"; 367.

39. Rossi, *Francis Bacon*, 13.

40. For Gouk, Bacon's reliance on the Aristotelian-Scholastic concept of the "species" to explain the conveyance of sound is an indication that he "was conceptually very distant from the mechanical philosophers who believed that sound is purely the impact of motion on the ear, perceived as sound by the listener"; see "Music in Francis Bacon's Natural Philosophy," 140.

41. "In Bacon's account of music's effects, the intellectual dimension of music and its effect on the rational soul is entirely ignored"; Gouk, "Music in Francis Bacon's Natural Philosophy," 143.

42. Gouk, "Music in Francis Bacon's Natural Philosophy," 140.

43. From the late Renaissance through the end of the seventeenth century, the idea that sound was a mystical entity (either imbedded in the spiritual nature of air or in the special qualities or species emitted by sounding bodies) with a reality outside of the mind was gradually discarded for the mechanical description of sound as matter in motion. For the history of this intellectual transition, see Finney, *Musical Backgrounds*, 139–58.

44. Ficino, *Opera Omnia*, 563, (*De Triplici Vita*, III., xxi) qtd. in Walker, *Spiritual and Demonic Magic*, 10.

45. Walker, *Spiritual and Demonic Magic*, 8.

46. Ficino, *Opera Omnia*, 1453, (*Comm. in Tim.*, c. xxviii) qtd. in Walker, *Spiritual and Demonic Magic*, 9.

47. On the tradition in acoustical literature that treats "sound as a material substance rather than an accident," and whose proponents include Priscian, Robert Grosseteste, and John of Salisbury, see Burnett, "Sound and its Perception," 67–69.

48. Kerrigan, Rumrich, and Fallon, eds., *The Complete Poetry*, 938.

49. According to Giambattista della Porta's scheme, magic has two kinds: "the one is infamous, and unhappie, because it hath to do with foul spirits, and consists of Inchantments and wicked Curiosity; and this is called Sorcery; an art which all learned and good men detest; neither is it able to yield any truth of Reason or Nature, but stands merely upon fancies and imaginations, such as vanish presently away, and leave nothing behind them.... The other Magick is natural; which all excellent wise men do admit and embrace, and worship with great applause; neither is there any thing more highly esteemed, or better thought of, by men of learning"; *Natural magick*, 1–2.

50. Shumaker explains, "traffic with ... unfriendly daemons was almost universally condemned as *goëtia*, or black magic. No matter what protestations of innocence might be made by the black magician, the consensus of informed opinion was that he entered into an implicit pact with the Devil, however unwittingly, just as the white witch or 'wise-woman' did. If so, he too was a witch. Usually, however, the witch made an explicit pact, agreeing to yield his soul ultimately to the Devil in exchange for extraordinary powers during his lifetime"; *The Occult Sciences*, 73.

51. *Natural magick*, 2.

52. Shumaker, *The Occult Sciences*, 121.

53. Shumaker, 132–33.

54. Agrippa, *Three books of occult philosophy*, 156.

55. Agrippa, 156.

56. Bacon, *Sylva Sylvarum*, 416. Bacon's idea for measuring the speed of sound was adapted by Mersenne and others; see Gouk, *Music, Science and Natural Magic*, 162.

57. Bacon, *Sylva Sylvarum* includes an entire section (experiments 124–37) on "[e]xperiments in consort touching production, conservation, and delation of sounds; and the office of the air therein"; 393–98. For Bacon's claim about the persistence of audible species in the air, see p. 432.

58. Bacon, 399, 418.

59. *Of the Dignity and Advancement of Learning*, 367.

60. "The Wisdom of Their Feet," 88–90.

61. "Song is the most powerful of all imitations because it reflects the intentions and moods of the spirit and stirs up those who hear it. Harmony is purer than matter and more like the sky than medicine, for it is warm, breathing air and, in a way, alive, composed of articulations and joints (*articulationibus artubusque*) like an animal, and possessed of a feeling and significance to which the sky will respond, as will the singer himself, especially if his nature is Phoebean (Apollo being the god of song)." Shumaker, *The Occult Sciences*, 133; paraphrase of Ficino, *Opera Omnia*, 1:558 (*De Triplici Vita*, 3.28).

62. *The Principles of Musik*, 114–15.

63. Bacon's discussion of this phenomenon occurs in century 3 of *Sylva Sylvarum*; 433. Gouk finds precedents for Bacon's description of sympathetic resonance in Pea-

cham's *Compleat Gentleman* (1622) and Porta's *Magia naturalis*; see Gouk, *Music, Science and Natural Magic*, 169.

64. Gouk, "Music in Francis Bacon's Natural Philosophy," 141.

65. All matter, according to Bacon, contains spiritus that enables it to perceive; see Gouk, "Music in Francis Bacon's Natural Philosophy," 141.

66. Lucretius, *On the Nature of Things*, 161.

67. *Of the Dignity and Advancement of Learning*, 367.

68. The madness Milton had in mind may have pertained to divinations of priests from Thracian or Phrygian religions and the cults of Cybele and Dionysus. "In these rites the participant was inspired by music and by dance to a religious frenzy in which he was possessed by the god or carried beyond himself"; see Finney, *Musical Backgrounds*, 51.

69. For a summary of the functions of animal spirits see Walker, "The Astral Body in Renaissance Medicine," 120.

70. Plotinus, Porphyry, Proclus, and Macrobius established a link between the spiritus, the soul's vehicle or container, and its divine origination and descent through the planetary spheres. Imagination is often construed as the "subtle body" that carries the soul out of the earthly body; see Klein, "Spirito Peregrino," 62–66.

71. Klein quotes from Dante's *Purgatory*, 9, 16–18, in the *Divine Comedy* (trans. Charles Singleton [Princeton, NJ, 1970–75]); see "Spirito Peregrino," 63.

72. See Walker, *Spiritual and Demonic Magic*, 5, 5n5.

73. Bacon, *Sylva Sylvarum*, 397.

74. Bacon, 597.

75. Bacon, 402.

3. THE POWER OF THE AIR IN MILTON'S EPIC POETRY

1. Shakespeare, *Merchant of Venice*, 4.1.180.

2. Ephesians 2:2 (KJV). For example, Death puns on pungent air when he says he won't "*err* / The way" to earth because he can smell the scent of carnage (10.266–68; emphasis added).

3. Edwards traces the influence of seventeenth-century natural history on Milton's epic (*Milton and the Natural World*). For Picciotto, *Paradise Lost* is symptomatic of the experimentalists' efforts to recover an Adamic sense of vision (*Labors of Innocence*, 400–507). Rogers examines the relations between *Paradise Lost* and the revolutionary language of mid-seventeenth-century monistic vitalism in *The Matter of Revolution*, 103–76. One exception to this trend is Marjara, whose well-informed discussion of meteorology in *Paradise Lost* demonstrates the interconnection of its natural systems (*Contemplation of Created Things*, 163–86).

4. Martin, *Renaissance Meteorology*, 14. For reasons the history of meteorology has suffered neglect, see 1–2 and 15–16.

5. Frytsche, *Meteorum, hoc est Impressionum Aerearum*, A6v; qtd. and translated in Martin, *Renaissance Meteorology*, 5.

6. *OED Online*, s.v. "atmosphere, n.1," accessed August 1, 2015. The word's first

known appearance in 1638 pertained to an extraterrestrial environment, the layer of air surrounding the moon; Wilkins, *The Discovery of a World in the Moone*, 138.

7. *OED Online*, 3rd ed., s.v. "air, n.1, 1.a.," accessed August 1, 2015.

8. Francis Bacon was an exponent of this theory. He writes in *Sylva Sylvarum* of the process of condensation, "We see it also in the effects of the cold of the middle region (as they call it) of the air; which produceth dews and rains" (348). For Milton's explication of the tripartite system, see Verity's note on Milton's concept of "middle air" ("Appendix D: *Paradise Lost*, 1.515–17," 674).

9. The narrator first identifies the "middle air" as the province of the Greek Olympian gods in the catalog of demons (1.516).

10. Treip argues that certain characterizations of the devil mix elements of realism with unrealism. One sign of this mixed allegorical mode is the too literal description of unreal or supernatural beings; another is the device of implication, or characterization by allusion to incongruous spaces and/or temporalities (*Allegorical Poetics and the Epic*, 239–47).

11. Ephesians 2:2, Luke 10:18, Colossians 2:15, and Ephesians 4:8 (KJV).

12. Milton mainly cites Ephesians 2:2 when he discusses Satan's God-given power over the spiritually dead or blind.

13. St. Jerome, "Ephesians 6:12," 258.

14. Arnold, *Ephesians: Power and Magic*.

15. Arnold, *Ephesians: Power and Magic*, 167–70.

16. Arnold, 60.

17. Augustine, *Books 8–11*, 343–47, 345.

18. Augustine explains further, "This name is said to have been derived from Juno, because in Greek Juno is called Hera, and therefore one or another of her sons, according to Greek mythology, was called Heros. Now the cryptic meaning of the myth is this. The air (*aer*) is counted as Juno's realm, and there, they would have it, the heroes dwell together with the demons" (*Books 8–11*, 345). According to one ancient tradition, Hera is associated with the element of air because in Greek her name is an anagram of "aer" (Kingsley, *Ancient Philosophy, Mystery, and Magic*, 15).

19. Augustine, *Books 8–11*, 345–47. For philosophical background on heroes and demons, see Algra, "Stoics on Souls and Demons," 71–96, especially 76 and 82–84.

20. Milton makes no distinction between the demons and the deities of the pagan world: "By falsities and lies the greatest part / Of mankind they corrupted to forsake / God their Creator" and "And devils to adore for deities" (1.367–69, 373).

21. St. Bede, St. Albertus Magnus, St. Thomas Aquinas, and St. Bonaventura all affirmed that weather phenomena could be affected by demons (White, *A History of the Warfare*, 1:336–38).

22. White gives a history of scripture-based meteorological explanation from Tertullian to the Schoolmen (1:323–31).

23. Martin, *Renaissance Meteorology*, 51–59.

24. White, *A History of the Warfare*, 1:323–36.

25. The corresponding passage in Latin: "deo permittente daemones possunt turbationem aeris inducere, ventos concitare, et facere ut ignis de caelo cadat" ("Expositio

super Job ad litteram," 5:3). For the English translation quoted above, see *The Literal Exposition of Job*, 71–90, 86. On the *Universae Naturae Theatrum*, see Blair, *The Theater of Nature*, 180–224, 184.

26. Blair, 145. Here, Blair paraphrases Bodin's discussion of violent weather's demonic source (for the original passage, see Bodin, *Universae Naturae Theatrum*, 160).

27. Athanasius, "The Life and Affairs of Our Holy Father Antony," 79.

28. Athanasius, 79.

29. Michelangelo, *The Torment of Saint Anthony*, ca. 1487–88, oil and tempera on panel; Kimbell Art Museum, Ft. Worth, Texas.

30. Hieronymus Bosch, *Triptych of the Temptation of St. Anthony*, 1500, oil painting on wood panels; housed at the Museu Nacional de Arte Antiga in Lisbon, Portugal; a print of the etching by Jacques Callot (*The Temptation of Saint Antony*, 1635) may be found in the William M. Ladd Collection, gift of Herschel V. Jones, 1916, at the Minneapolis Institute of Art.

31. Palma Vecchio, *Sea Storm*, ca. 1508–28, oil on canvas; housed at the Galleria dell' Accademia, in Venice, Italy. On the difficulty of precisely dating the painting, see Sohm, "Palma Vecchio's *Sea Storm*," 89.

32. Sohm, "Palma Vecchio's *Sea Storm*," 85.

33. Luther, *Table Talk*, 82.

34. "[Q]uamvis enim materia corporalis non oboediat ad nutum angelis neque bonis neque malis ad susceptionem formarum sed soli creatori deo, tamen ad motum localem natura corporea nata est spirituali naturae oboedire" (Aquinas, "Expositio super Job ad litteram," 3); and Aquinas, *The Literal Exposition of Job*, 77.

35. "Selections from *The Second Defense of the English People*," in Kerrigan, Rumrich, and Fallon, *The Complete Poetry*, 1069–110, 1079.

36. Martin, *Renaissance Meteorology*, 42.

37. Martin, 42–43.

38. See Aristotle, *De Generatione et Corruptione*, 41–43. In *Meteorologica*, 11, 23, Aristotle reiterates the point that all of the elements are transformable into each other.

39. This was his initial position, although later in life Augustine decided that the demons dwell in a subterranean hell (Fitzgerald, ed., *Augustine through the Ages*, 247).

40. Augustine, *Books 1–6*, 84.

41. Augustine, 83.

42. Lombard, *The Sentences, Book 2*, 34.

43. Augustine, *Books 1–6*, 84.

44. Consider the central role exhalations play in this explanation of earthquakes, for example: "[T]here must be exhalation both from moist and dry, and earthquakes are a necessary result of the existence of these exhalations. For the earth is in itself dry but contains much moisture because of the rain that falls on it; with the result that when it is heated by the sun and its own internal fire, a considerable amount of wind is generated both outside it and inside" (Aristotle, *Meteorologica*, 205). For a summary of Aristotelian exhalation theory, see Taub, *Ancient Meteorology*, 88–91 and 114.

45. Martin, "Causation in Descartes' *Les Météores*," 217–36, 227.

46. Aristotle, *Meteorologica*, 167.

47. Aristotle, *Meteorologica*, 29–35, 69–75, and 223–31. For a summary of the processes by which meteorological phenomena such as shooting stars, thunder, and lightning are formed from dry exhalation, see Taub, *Ancient Meteorology*, 90.

48. Deuteronomy 8:3 (KJV).

49. Recall that Satan is sometimes depicted as "nitrous" (4.815), thunderous (1.601), fiery (2.1013), and even sunny (1.594–96 and 2.492–95).

50. On the operation of the dry exhalation within the earth, see Taub, *Ancient Meteorology*, 99–100.

4. "HOW CAM'ST THOU SPEAKABLE OF MUTE"

1. On the basis of "the single effect of grandeur of sound," Eliot finds that there is "nothing finer in poetry" than what Milton achieves; "Milton I," 158, 164, originally published in 1936 as "A Note on the Verse of John Milton." Eliot means "best" in a both a personal and a relative sense (i.e., compared to other poets).

2. The downside of what Eliot dubs Milton's "rhetorical style" is "that a dislocation takes place, through the hypertrophy of the auditory imagination at the expense of the visual and tactile, so that the inner meaning is separated from the surface, and tends to become something occult, or at least without effect upon the reader until fully understood"; "Milton I," 162.

3. Leonard's ample account of the "Milton Controversy" of the twentieth century and its roots in early Milton criticism, traces among other things fluctuating opinions of Milton's sonorous style. Leonard identifies Daniel Webb's *Remarks on the Beauties of Poetry* (1762) as the first work of criticism to connect the sound of Milton's blank verse to organ music. In the nineteenth century, Leigh Hunt, Tennyson, and James Russell Lowell used an organ metaphor to discuss Milton's musical virtuosity. However, after Ezra Pound critiqued Milton for using highly latinized language, "the 'sonority' that had hitherto been seen as a virtue will be seen as a vice, and eulogies of Milton's 'organ music' will only harden the opposition." *Faithful Labourers*, 1:59–265, 174. For Leavis's disparagement of the pompous sound of Milton's poetry, see "Milton's Verse," 126.

4. Berley reads *Paradise Lost* as deeply influenced by speculative music in its cultivation of a poetic of striving for heavenly song (*After the Heavenly Tune*, 180–205); Buhler, "Counterpoint and Controversy," argues that Milton's depiction of polyphonic music reinforces arguments made by reformers against contrapuntal music; Minear's analysis of *Paradise Lost* explores the poem's evaluation of verbal and nonverbal forms of music and its self-representation by these standards (*Reverberating Song*, 227–56); Ortiz contends that in *Comus*, Milton employs music in protest of Reformist strictures that would eliminate figuration (see *Shakespeare and the Politics of Music*, 213–42).

5. "The tendency has been to think in 'logocentric' terms, as if there were language, and then as a separate and unrelated category, non-linguistic sound. . . . What an 'acoustic approach' can bring to this discussion, though, is the observation that *Paradise Lost* confounds language and music with all other sorts of noise into a continuum." Steggle, "*Paradise Lost* and the Acoustics of Hell," 12. Sherry argues that the

sound of the epic is "material and animate," like every other part of Milton's cosmos and that its substance is endued with spirit, or animated, by the poet's creative process; see "Milton, Materialism, and the Sound," 220.

6. For example, Buhler connects "an alertness toward music's affective powers" in Milton's poetry to the prevalent notion among reformers that "polyphony's appeal to the senses cannot be balanced by language's appeal to the rational soul." "Counterpoint and Controversy," 19–20.

7. Sherry, "Milton, Materialism, and the Sound," 224.

8. In a rare analysis of the air in *Paradise Lost*, Jayne Elizabeth Lewis connects its fall with the rise of what she calls "literary atmosphere" in the long eighteenth century. For Lewis, Satan's influence on the atmosphere of Paradise renders it analogous to the literary medium of Milton's poem: the fallen air becomes visible, material, and limitedly expressive. This reading, though valuable for its appreciation of literary atmospheric resonance, discounts the consistently acoustical nature of Satan's pneumatic incursions and ignores their meteorological genesis. *Air's Appearance*, 36–40, 53–54.

9. Minear, *Reverberating Song*, 234.

10. "Milton's 'Unoriginal' Voice," 159.

11. "Forge bellows" were used to supply air to pipe organs, replacing more fragile animal bladders with a technology originally intended for stoking fires; see Bush and Kassel, *The Organ*, 63.

12. *OED Online*, s.v. "assay, v.4," accessed Dec. 1, 2015.

13. *OED Online*, s.v. "assay, v.6," accessed Dec. 1, 2015.

14. "Satan attempts in *Paradise Lost* to reach either directly to the organ of fancy, highest of the powers which he could subject to his rule, or alternatively (4.804) to the animal spirits which were the source of sense data and which would retain somewhat past experiences which he could mold to his own purposes." Hunter, "Eve's Demonic Dream," 263.

15. "Minutes of the Life of Mr. John Milton," in Kerrigan, Rumrich, and Fallon, *The Complete Poetry*, xxiii–xxx, xxvii.

16. Spaeth, *Milton's Knowledge of Music*, 16, 16n4, 22.

17. "Organs first in France. The ambassadors of the Greek Emperor Constantine brought King Pepin some organs, which had never before been seen in France."

18. *OED Online*, 3rd ed., s.v. "organ, n.1.," accessed December 1, 2015.

19. "The moist exhalation, then, is the material of the metals. Along with portions of the dry exhalation it is trapped underground, where it condenses, particularly if it comes into contact with rocks, and then hardens, probably through cold. . . . Because the metals contain earthy matter, they cannot revert to water, and for the same reason they are, with the exception of pure gold, affected by fire" (paraphrase of Aristotle's theory from *Meteorologica* in Eichholz, "Aristotle's Theory of the Formation of Metals and Minerals," 143).

20. There was some disagreement about how metals came to be. While many Renaissance theorists thought that metals were derived from some combination of mercury, sulfur, and nitre, others identified exhalation as their primary ingredient. Marjara explains that the main ingredients thought to be involved in generating metals changed

over time. In the early Renaissance, alchemists believed that sulfur and mercury were responsible for the formation of metals; later, in the seventeenth century, sulfur and nitre were the preferred combination; see *The Contemplation of Created Things*, 172–78.

21. Heninger, *A Handbook of Renaissance Meteorology*, 3.

22. Heninger, 4, 147–48.

23. For a concise explanation of Aristotelian exhalation theory, the predominant meteorological model throughout the Renaissance, see Heninger, 9.

24. Pietro d'Abano links sonority to the inward structure of metal. Similar discussions appear in the pseudo-Aristotelian *Problemata* and the *Quaestiones* of Nicolaus Peripateticus; see Burnett, "Sound and its Perception," 51n70.

25. Burnett, "Sound and its Perception," 51.

26. "Et ideo ista sonora sunt vehementer et dui retinent sonum, eo quod aëre plena sunt, quae trementia ex ictu forti continue a se expellunt." *De anima*, 2.3.17.27–29, 124; translation by Burnett, "Sound and its Perception," 51.

27. This opposition derives from the antagonism between the *kithara* and the *aulos* in the mythological contests between Apollo and Marsyas and Apollo and Pan, in which Phoebus's string music always triumphs over the satyrs' pipes. Plato denounces the playing of *aulos* as devoted only to pleasure in *Republic* 3. For the classical background and further discussion of historical symbolism of string and wind instruments, see Winternitz, *Musical Instruments and Their Symbolism*, 150–65.

28. For instance, on the first Sabbath "the solemn pipe, / And dulcimer, all organs of sweet stop" join the voices of heaven in a hymn of Creation (7.595–96).

29. *The Books on Instruments*, 391.

30. Wilkins, *Mathematical Magick*, 10. For the distinction between divine, natural, and artificial sciences, see 1–2.

31. Hunt, "An Organ in the House (Introductory Article)," 159.

32. Wilkins, *Mathematical Magick*, A4r.

33. Wilkins, 3–4.

34. The inconceivably swift construction of Pandaemonium outfaces the ingenuity of man's greatest building feats (1.692–99).

35. See Schott, *Magia Universalis*, 2:300–302; *Mechanica Hydraulico-Pneumatica*, 383–440; Kircher, *Musurgia Universalis*, 2:330–35.

36. For example, Kircher addresses hydraulic organs in *Musurgia Universalis* in a section on "Magia Phonocamptica" (the magic of reflected sounds and artificially created echoes). On Kircher's and Schott's contribution to acoustics as a pursuit of natural magic, see Gouk, *Music, Science and Natura Magic*, 101–9.

37. Wilkins, *Mathematical Magick*, 149. See also Gouk, *Music, Science and Natural Magic*, 168.

38. Wilkins discusses other automata that could make meaningful sounds, giving the oft-cited examples of medieval magician Roger Bacon's "brazen head" and the speaking statue of Albertus Magnus. Rather than affirming or denying such legends, Wilkins ponders the practical means by which machines might be made to speak (*Mathematical Magick*, 177–78). A few years after *Mathematical Magick* was published, John Evelyn recorded in his diary that Wilkins, whom he visited at Wadham College, had invented a speaking statue of his own (entry for July 13, 1654, in *The Diary of John*

Evelyn, 298). Developing alternative methods of communication was a lifelong fascination. Wilkins devoted an earlier work, titled *Mercury, or, The secret and swift messenger: Shewing, how a man may with privacy and speed communicate his thoughts to a friend at any distance* (1641), to recording various methods of encrypting or concealing messages, and his *magnum opus, An Essay towards a Real Character, and a Philosophical Language* (1668), proposed a full-fledged universal language. For more on the tradition of artificially animated heads see Dickson, *Valentine and Orson*, 200–216.

39. Walters, *Church Bells of England*, 256–57.

40. "Et cum melodia illius auribus insonuerit populorum, crescat in eis devotio fidei; procul pelantur omnes insidiæ inimici, fragor grandinum, procella turbinum, impetus tempestatum; temperentur infesta tonitrua; ventorum flabra fiant salubriter, acmoderate suspense; prosternat aëreas potestates dextera tuæ virtutis; ut hoc audietes tintinnabulum contremiscant, & fugiant ante sanctæ crucis Filij tui in eo depictum vexillum"; *Pontificale Romanum Clementis VIII*, 299. See also Lynch, *Right of the Blessing of a Bell*, 23.

41. White, *A History of the Warfare*, 1:345.

42. Walters, *Church Bells of England*, 263; see also the entry on the parish of Gulval in Polsue, *A Complete Parochial History*, 115–17. For the dates of the bell tower construction see Cox, *Cornwall*, 118.

43. Walters cites several church records that provide for bell ringing in the time of dangerous storms; see *Church Bells of England*, 262.

44. Longfellow, *The Golden Legend*, 1–2.

45. Lynch, *Right of the Blessing of a Bell*, 22.

46. Smith and Cheetham, eds., *Dictionary of Christian Antiquities*, 1:185.

47. In ancient Rome and Greece livestock, warriors, and the deceased were all adorned with bells to ward off predators and evil-doing spirits. Coleman, *Bells*, 26–28. On *tintinnulabum*, the protective symbol of the phallus was often displayed along with bells. An example of these protective chimes appears in Simons, *The Sex of Men*, 55–56.

48. Gouk, "Transforming Matter, Refining the Spirit," 147–49, 155–56.

49. "Transforming Matter, Refining the Spirit," 155.

50. Gouk observes that Khunrath and Agrippa both hold sacred harmony as a deterrent to evil. She translates the inscription in an engraving titled "Lab-Oratorium" from Khunrath's *Amphitheatrum sapientiae aeternae* (1595) depicting an assortment of stringed instruments in an alchemist's laboratory as follows: "Sacred Music is the dispeller of sadness and evil spirits, because the Spirit [SPIRITUS] of Jehovah gladly sings in a heart filled with pious joy"; see "Transforming Matter," 151.

51. Gouk, "Transforming Matter, Refining the Spirit," 146–47, 149, 155–56.

52. Michael Praetorius's treatise on instruments is a classic example (*Syntagma Musicum II: De Organographia*). *De Organographia* was first published in Wolfenbüttel in 1618 and 1619.

53. Gouk, *Music, Science and Natural Magic*, 176.

54. Gouk, 176.

55. Mersenne, *The Books on Instruments*, 294.

56. Mersenne, 383.

57. Mersenne, 383.

58. Mersenne, 383.

59. Mersenne, 294.

60. Milton uses "wind" in a similarly witty vein to characterize the pneumatic conveyance of sound in *L'Allegro* ("Lap me in soft Lydian airs, / . . . / In notes, with many a winding bout" [136–39]) and in *Comus* ("Wind me into the easie-hearted man" [163]). The word evokes the airy quality of sound that enables it to wiggle into physically and psychologically confined spaces.

61. *Sylva Sylvarum*, 391. See also, for example, 404, 422.

62. A demon in the interior of Hieronymus Bosch's *Triptych of Temptation of St. Anthony* (ca. 1500, Museu Nacional de Arte Antiga, Lisbon) is shown with a pommer (shawm) nose pipe. Others carry a harp, lute, and an organistrum. Nose pipes appear in at least four other scenes of hell by Bosch and his followers, and anus pipes in at least one; see Planer, "Damned Music," 340–41, tables 2 and 3. See also Jacques Callot, *The Temptation of Saint Antony*, 1635, discussed on p. 112.

63. Some of the embodied instruments in Bosch's works have been interpreted as symbols of punishment: "the instruments torment the damned directly by imprisoning them . . . ; penetrating their orifices; or ensnaring them"; Planer, "Damned Music," 343.

64. Satan predicts that on hearing (and feeling) the effect of his cannons the angels will "fear we have disarmed / The thunderer of his only dreaded bolt" (6.490–91).

65. The sack of the bagpipe is typically "made of the skin or bladder of an animal, usually a goat or a sheep." Winternitz, *Musical Instruments and Their Symbolism*, 67.

66. *OED Online*, s.v. "embowel, v.3.a.," accessed December 15, 2015.

67. Just after this observation about cannon fire, Bacon proposes a method for measuring sound delay by standing at a great distance from a bell in a steeple and recording the difference between the moment it is observed as struck and the moment its peal is heard. See, *Sylva Sylvarum*, 416.

68. Hunt, *Origins in Acoustics*, 85. For information on the "blast-timing" experiments of Mersenne, Gassendi, and the scholars of the Accademia, see 99–104.

69. Campbell, Greated, and Myers, eds., *Musical Instruments*, 160. Traditionally, the inventor is given as Edmé Guillaume of Auxerre, France, and the date of invention, 1590.

70. Campbell, Greated, and Myers, 478. The nominal size of a typical serpent is 8-foot C, which makes the equivalent cone length 2.65m or about 8 foot 8 inches.

71. Mersenne, *The Books on Instruments*, 350. See also, Campbell, Greated, and Myers, *Musical Instruments*, 159–60.

72. Campbell, Greated, and Myers, *Musical Instruments*, 160. Mersenne indicates that the crook, which he calls a tube, may be "ivory, horn, silver, or tin"; see *The Books on Instruments*, 353.

73. Mersenne, *The Books on Instruments*, 353.

74. Mersenne, 353.

75. Kircher, 6.4.505.

76. Campbell, Greated, and Myers, *Musical Instruments*, 393.

77. S.v. "Vox Humana/Voix Humaine/Voce Umana," Bush and Kassel, *The Organ*, 612.

78. "Milton's 'Unoriginal' Voice," 173.

79. Campbell, Greated, and Myers, *Musical Instruments*, 387, 389.
80. *OED Online*, s.v. "tongue, n.14.c.," accessed December 15, 2015.
81. Campbell, Greated, and Myers, *Musical Instruments*, 390.
82. Helkiah Crooke in *A Description of the Body of Man* (1616) includes the windpipe or "weazon" in his account of the organs that produce the human voice. In the *The French Academie* (1618), Pierre de La Primaudaye refers to the human vocal apparatus as a portative organ. See Smith, *The Acoustic World*, 4.
83. *OED Online*, s.v. "impulse, n.3.a.," accessed December 15, 2015.
84. *Paradise Lost. A Poem in Twelve Books*.
85. *The Acoustic World*, 125.
86. See Smith, *The Acoustic World*, 125, on the organic names of early modern printing house technology.

5. MILTON AND THE BAROMETER

1. In *Areopagitica*, Milton reports that he "found and visited the famous Galileo grown old, a prisner to the Inquisition, for thinking in Astronomy otherwise then the Franciscan and Dominican licencers thought" (*CPW* 2:538). Given that no corroborating accounts of the visit are known, critics have long debated whether the meeting actually took place. Most biographers take Milton at his word. For a recent example of scholarship that doubts the veracity of Milton's claim, see Butler, "Milton's Meeting with Galileo." For a defense of the meeting and a rigorous demonstration of Galileo's influence on Milton's astronomy, see Nicholson, "Milton and the Telescope." *Paradise Lost* alludes to Galileo and his instrument at least three times at 5.261–63, 1.287–91, and 3.588–90.
2. Demonstrating a marked difference between Milton's poetics of space before and after the trip, Nicholson argues that he must have looked through an actual telescope; "Milton and the Telescope," 10.
3. For other passages where Milton's "unprecedented aesthetics of space" reflects Galileo's new astronomy, see Kerrigan, Rumrich, and Fallon, *The Complete Poetry*, 258.
4. The epithet refers to Galileo.
5. "La scoperta del Barometro insomma fece cambiare l'aspetto della fisica, come il Telescopio quello dell' astronomia, la circolazione del sangue quello della medicina, la Pila del Volta quello della fisica molecolare"; Antinori, *Notizie istoriche relative all'Accademia del Cimento*, 27, qtd. and trans. in Gliozzi, "Torricelli, Evangelista," 439.
6. Middleton, *The History of the Barometer*, 3–82, provides a chronological account of these pneumatic experiments. One of the original experiments, Evangelista Torricelli's demonstration with an inverted tube of mercury, had the greatest impact and was repeated throughout Europe. The arrival of Galileo's *Discorsi e dimostrazioni matematiche* (*Two New Sciences*) (1638) in Rome in December of 1638, around the time of Milton's first visit to that city, may have sparked an earlier experiment of a similar kind. In response to Galileo's assertion about the behavior of siphons, a group of men, led by Gasparo Berti and including Rafael Magiotti and Father Athanasius Kircher, built a contraption that anticipates Torricelli's barometer. The precise dates of both Berti's and Torricelli's experiments are not known, but Middleton estimates the former

took place in 1641 and the latter not long before Torricelli describes it in a letter dated June 11, 1644 (*The History of the Barometer*, 20, 30).

7. Trubowitz, "Milton's Chaos and the Vacuist–Plenist Controversy," is a notable exception for engaging closely with a central debate in pneumatics, particularly its influence on the illusion of vacuity in Milton's Chaos, which Trubowitz shows is nonetheless part of a fundamentally plenist universe. For Trubowitz, Milton's engagement with the idea of air pressure amounts to a swipe at the experimentalists whose measurements with the Torricellian tube took them, like Nimrod and seventeenth-century imperialists, above the mountaintops to breathe thin air. Joanna Picciotto also touches on parallels between Milton's epics and the goals of the experimentalists, some of whom were centrally involved in the history of pneumatics. But her comparison does not directly juxtapose Milton's views with their findings regarding the air. Instead, Picciotto is concerned with the poetry's spectatorial ambitions and thus, its functional likeness to key optical instruments of the experimental science. See *Labors of Innocence*, 400–505.

8. For example, Cesare Cremonini, a professor at Padua, rejected Galileo's observations with the telescope, arguing that he had proved nothing so long as he could not provide a philosophical synthesis of the cosmos. See Heilbron, *Galileo*, 195–96.

9. See, for example, the usage of Pierre Petit, the first person outside of Italy to successfully reproduce Torricelli's device, in a published letter addressed to Pierre Chanut, the ambassador to Sweden: "C'est de *l'experience du Toricelli* [sic] touchant le vuide, dont ie vous veux entretenir si vos affaires plus serieuses vous en peuuent donner le loisir" (emphasis added; *Observation touchant le vuide faite pour la première fois en France* [. . .] [Paris, 1647] qtd. in Middleton, *The History of the Barometer*, 38).

10. *OED Online*, 2nd ed., s.v. "barometer, n.a.," accessed May 1, 2019. See also the gloss on "barometer n." in *OED Online*, 3rd ed., s.v. "-meter, comb. form2," accessed May 1, 2019. Middleton cites an earlier instance of "barometer" in Robert Boyle's *New Experiments and Observations touching Cold* (London, 1665), 27; *The History of the Barometer*, 71.

11. Giles, "Pressure, Surface," 586.

12. Torricelli's intention was to create a device that illustrates fluctuations in atmospheric pressure and thus possible changes in the weather; Middleton, *The History of the Barometer*, 24–25, 29; Howard, *The History of Meteorology to 1800*, 69.

13. Camuffo et al., "The earliest daily barometric pressure readings in Italy," 337.

14. John Locke, for example, gathered regular observations in 1692–1703; see Camuffo, 344. Boyle calls for greater organization among barometer users in "Some Observations and Directions about the Barometer."

15. Galileo passed away on January 8, 1642. Gliozzi, "Torricelli, Evangelista," 433–34.

16. Galileo had already died by the time Torricelli conducted the experiment, but his assistant Vincenzo Viviani had taken notes at Galileo's direction on the difference in height at which quicksilver, water, and other liquids reach their breaking point in response to the action of a pump. Torricelli's use of mercury instead of water allowed for the convenient size of the barometric apparatus; see Middleton, *The History of the Barometer*, 20.

17. Galileo fails to comprehend that air pressure is the force that creates suction. For his explanations of the attractive force of vacuum and the functioning of water pumps that rely on suction, see *Two New Sciences*, 19–22, 24–25 (*Le Opere di Galileo Galilei*, 8.59–61, 64).

18. 18 cubits, or *braccia*, in the Tuscan dialect. For Baliani's suggestions, see Middleton, *The History of the Barometer*, 9.

19. Galileo, *Two New Sciences*, 24–25.

20. Although the precise date of the experiment is not known, Middleton estimates it took place in 1641; *The History of the Barometer*, 20.

21. For a seventeenth-century account of Berti's experiment, see Emmanuel Maignan's description in *Cursus philosophicus concinnatus ex notissimis cuique principiis, ac praesertim quod res physicas instauratas ex lege Naturae sensatis experimentis passim comprobata*, 4 vols. (Toulouse, 1653), 1925–1936, qtd. and translated in Middleton, *The History of the Barometer*, 10–15.

22. Middleton cites a passage from a letter from Magiotti to Marin Mersenne in 1648: "Berti believed he could convince Galileo with this experiment"; MS 7049, letter CXXVII, Nationalbibliotek in Vienna, qtd. and translated in *The History of the Barometer*, 16. What Berti meant to convince him of is unclear. Baliani had tried to persuade Galileo that weight of the air, not the vacuum *per se*, causes the elevation of water in a pump to a certain height. But Gliozzi writes that the purpose of Berti's experiment was to show, "that the water in suction pumps rose to more than eighteen *braccia*"; "Torricelli, Evangelista," 438.

23. Galileo's interpretation of the behavior of suction pumps reflected his controversial belief that matter contains *vacuum disseminatum*, or void that is in a subdivided and distributed form. See Webster, "The Discovery of Boyle's Law," 444.

24. Aristotle, *On the Heavens*, 353.

25. Aristotle, 353.

26. Aristotle, 355.

27. Simplicius in his commentary on Aristotle's *De Caelo* discusses both Ptolemy's and his own repetition of Aristotle's experiment. Ptolemy's *On Weights* is now lost. See Sambursky, *The Physical World*, 80.

28. Sambursky, *The Physical World*, 78–80.

29. Simplicius qtd. in Sambursky, 81.

30. "[I]f natural pull is a striving towards the proper place, objects which are there should not strive towards it nor pull in that direction, since they are already there. He who has had his fill does not reach for food," Simplicius qtd. in Sambursky, *The Physical World*, 80.

31. Desaguliers, *A Course of Experimental Philosophy*, 2:96.

32. *Two New Sciences*, 80, 84 (*Le Opere di Galileo Galilei*, 8.121, 125).

33. The statement is from an anonymous essay on the elements dating to ca. 1150–75. The author(s) seem to have known Aristotle's *Physics*. From the text of R. C. Dales, "Anonymi De Elementis," *Isis*, 56 (1965): 174–89, translated and abridged in Dales, *The Scientific Achievement*, 52–60, 59.

34. *Two New Sciences*, 82 (*Le Opere di Galileo Galilei*, 8.123), emphasis added.

35. See Beeckman's personal journal (*Journal tenu par Isaac Beeckman de 1604 à*

1634, 1:36) and Descartes' letter dated June 2, 1631 (René Descartes, *Oeuvres*, eds. Chas. Adam et Paul Tannery [Paris, 1897–1911], vol. 1 [1897], 205–8); both are excerpted and translated in Middleton, *The History of the Barometer*, 6–8.

36. *Oeuvres*, 1:205–8, qtd. in Middleton, *The History of the Barometer*, 7.

37. Isaac Beeckman, 24 December 1620–19 February 1621 in vol. 2 of *Journal tenu par Isaac Beeckman de 1604 à 1634*, qtd. and translated in Webster, "The Discovery of Boyle's Law," 445, emphasis added.

38. Pecquet, *New anatomical experiments*, 90, 108.

39. Beeckman was visited by Marin Mersenne, Rene Descartes, and Pierre Gassendi, and Mersenne and Descartes read his journal. See Hooykaas, "Beeckman, Isaac."

40. Milton's line, "Coelum non animam muto dum trans mare curro," adapts Horace's *Epistle* 11.27. Translation by Lewalski, *The Life of John Milton*, 108.

41. An approximate date of the experiment based on Torricelli's June 11, 1644, letter to Michelangelo Ricci (1619–1692); see Middleton, *The History of the Barometer*, 29–30.

42. See note 16. After Galileo's death, Viviani became a staunch promoter of his legacy. Failing to erect a public monument, Viviani built memorial facades outside of his palace as a grand tribute to his teacher. See Stephano Gattei, "Galileo's legacy."

43. A Frenchman named Du Verdus, who had been shown Torricelli's letters to Ricci relating to the famous experiment, sent passages from them to Mersenne; see Middleton, *The History of the Barometer*, 35–37.

44. Mersenne himself went to Rome and Florence in 1644–45 where Torricelli showed him the experiment. Middleton believes that the extracts copied by Du Verdus from Torricelli's letters informed the first experiment of the kind conducted in France by Pierre Petit and Blaise Pascal in 1646 (*The History of the Barometer*, 37–39). See Torricelli's letter to Ricci (June 11, 1644) (G. Loria and C. Vasura, eds., *Opere di Evangelista Torricelli*, vol. 3 [Faenza: G. Montanari, 1919], 186–88), which is quoted in full in *The History of the Barometer*, 23–24. Hereafter, I quote from Middleton's translation.

45. Middleton, *The History of the Barometer*, 23. Webster, "The Discovery of Boyle's Law," 446.

46. Middleton, *The History of the Barometer*, 23–24.

47. Middleton, 24.

48. Middleton, 24.

49. Middleton, 23; emphasis added.

50. In the first sense defined by Cohen, *The Newtonian Revolution*, 159–60.

51. The term "fluid-heaven theory" is original to Lattis, *Between Copernicus and Galileo*, 94. See 94–102 for the history of this school of thought.

52. Three writers in the tradition, Andalo di Negro (ca. 1270–1340), Giovanni Pontano (1426–1503), and Robert Bellarmine (1542–1621), used this simile or some variation of it, and critics of the fluid-heaven theory, such as the Jesuit philosopher Christoph Clavius, coopted the analogy in their rebuttals; Lattis, 98. On the fluid-heaven cosmology as concomitant to the acceptance of the Tychonic system, see 211–16.

53. [Oldenburg and Beale], "*Observations Continued upon the* Barometer," 165.

54. The opening beneath heaven from which a passage extends to earth is bounded like an ocean's shore to shut out the darkness of chaos (3.538–39).

55. See Shank, "What Exactly Was Torricelli's 'Barometer'?" 166, for an explanation of the means through which an abridged form of Torricelli's letter, compiled by François du Verdus, passed to Mersenne and throughout France.

56. Dati shares the news of his appointment with Milton in a letter dated December 4, 1648; "[Carlo Dati to Milton]," *CPW* 2:774–75. On Dati's involvement with the Accademia del Cimento, see Middleton, *The Experimenters*, 27, 31.

57. On Dati's tribute to Milton, see Lewalski, *The Life of John Milton*, 102–3.

58. Lewalski, 210, 223.

59. Webster, "The Discovery of Boyle's Law," 446.

60. For a discussion of Pascal and Perrier's Puy du Dôme experiment, which showed the effect of altitude on a barometer set up on a mountain in Auvergne, see Cajori, *A History of Physics*, 66. For Roberval's observations of water, air, and a carp bladder introduced into the vacuum space inside a barometer, see Webster, "The Discovery of Boyle's Law," 448–50.

61. "Hartlib to Boyle, 9 May 1648," in *The Correspondence of Robert Boyle*, 1:65–66. The above quotation is taken from an excerpt of a letter by Charles Cavendish (1591–1654) to William Petty (1623–87) included in Hartlib's letter.

62. On Haak's connection to Milton through their mutual work for the Council of State, and/or friendship with Samuel Hartlib; his spearheading the "1645 Group"; and their experiments with the tubes of mercury, see Barnett, *Theodore Haak*, 75–82, 87–88, 91.

63. Haak to Mersenne, letter of 24 March 1648, qtd. in Boas, "The Establishment of the Mechanical Philosophy," 419.

64. Barnett, *Theodore Haak*, 91.

65. Barnett, 87.

66. Wallis, "Account of some Passages of his own Life [1696–97]," 1.clxiii. See also Lyons, *The Royal Society, 1660–1940*, 9.

67. One is the aforementioned 1653 translation of Pecquet's *New anatomical experiments* (see note 38). The other is Charleton, *Physiologia Epicuro-Gassendo-Charltoniana*.

68. On dating the construction of the pump, see Wilson, "On the Early History of the Air-Pump," 6. The ability to isolate the Torricellian tube of mercury in the receiver (the seventeenth experiment presented in the *New Experiments Physico-Mechanical*) was arguably the chief reason Boyle constructed the air pump; see Shapin and Schaffer, *Leviathan and the Air-Pump*, 40.

69. Wren to Brouncker, 30 July/9 August 1663, in Thomas Birch, *The History of the Royal Society of London for the Improving of Natural Knowledge, from its First Rise*, vol. 1 (London, 1756–57), 288, qtd. in Shapin and Schaffer, *Leviathan and the Air-Pump*, 31.

70. Ibid.

71. Rumrich, *Matter of Glory*. See especially "Glory in the Old Testament" (14–25).

72. See, for example, Hainsworth, "Ups and Downs in *Paradise Lost.*"

73. "Elater" is Pecquet's coinage, derived from the transliterated Greek word *elasticus* (ελαστικός) meaning propulsive or impulsive. For the transmission of Pecquet's concept of the elatery of air into Boyle and other English authors and the evolution of the word elasticity, see Webster, "The Discovery of Boyle's Law," 451–54.

74. Pecquet, *New anatomical experiments*, 90; emphasis added. Robert Boyle follows

this passage very closely in his description of the spring of the air in *New Experiments Physico-Mechanicall, Touching The Spring of the Air:* "[O]ur Air either consists of, or at least abounds with, parts of such a nature, that in case they be bent or compress'd by the weight of the incumbent part of the Atmosphere, or by any other Body, they do endeavor, as much as in them lies, to free themselves from that pressure, by bearing against the contiguous Bodies that keep them bent"; *The Works of Robert Boyle,* 1:165, experiment 17.

75. The pump was built by Robert Hooke and adapted from an earlier design by Otto Von Guericke (1654). Boyle and Hooke's "Pneumatical Engine" could approximate the conditions of a vacuum in its large glass receiver or globe, which had a capacity of 1732 cubic inches, or about 7.5 liquid gallons. In other words, there was room enough inside to hold a Torricellian tube sitting in a dish of mercury. See Wilson, "On the Early History of the Air-Pump," 8.

76. This is the term Boyle and others use to refer to the liquid in the vessel next to and below the column of mercury in the Torricellian tube; see *The Works of Robert Boyle,* 1:192.

77. "Vacuity in Vacuity" is Pecquet's phrase; see *New anatomical experiments,* 104.

78. Boyle, *The Works of Robert Boyle,* 1:193.

79. Boyle, 1:193.

80. I am grateful to John Leonard for pointing out Harding's insight to me. See *The Club of Hercules,* 54. For further discussion of the Virgilian allusion, see Leonard, *Faithful Labourers,* 1:467.

81. *OED Online,* s.v. "incumbent, adj. 1.a.," accessed June 6, 2019.

82. *OED Online,* s.v. "incumbent, adj. 2.a.," accessed June 6, 2019.

83. "1660 R. Boyle *New Exper. Physico-mechanicall,* ii, 38. The Atmosphere incumbent upon the upper part of the same key or stopple"; *OED Online,* s.v. "incumbent, adj. 2.a." (see note 82).

84. "Aqua suctu sublata non attrahiturvi vacui, sed ab aere incumbente in locum vacuum impellitur"; see "Corollaria [6 September 1618]," in *Journal tenu par Isaac Beeckman de 1604 à 1634,* 4:44.

85. *New anatomical experiments,* 90.

86. *Physiologica Epicuro-Gassendo-Charltoniana,* 48.

87. Charleton, *Physiologica Epicuro-Gassendo-Charltoniana,* 47.

88. *The Works of Robert Boyle,* 1:192–93.

89. See Flannagan, ed., *The Riverside Milton,* 86n.

90. See 10.529.

91. Edmund Spenser, *The Faerie Queene,* bk. 1, canto 11, st. 18, qtd. in Leonard, ed., *Paradise Lost,* 296n227.

92. "The Language," *Spectator* 285 (January 26, 1712) 266, qtd. in Leonard, *Faithful Labourers,* 1:19.

93. "A Relation of some Mercurial *Observations,*" 155.

94. [Oldenburg and Beale], "*Observations Continued,*" 163; Boyle, "Of a New kind of Baroscope," 237.

95. See, for example, in *Physiologica Epicuro-Gassendo-Charltoniana:* "Water infused into the Tube doth also descend to the point of Æquipondium, and stops to the altitude

of 32 Feet, no more, no less" (52); and *"Why the Æquilibrium of these two opposite Forces, is constant to the certain præcise altitude of 27 digits?"* (50).

96. *Contemplation of Created Things*, 207.

97. The ancient Greek mathematician and expositor of the principles of levers was supposed to have said, famously: "Give me a place to stand and I will move the earth."

98. See *Mathematicall Magick*, 81. For another vignette of the Archimedean boast depicting a man using a balance to move the suspended globe, see Pierre Varignon, *Projet d'une nouvelle méchanique, avec un Examen de l'opinion de M. Borelli sur les propriétez des poids suspendus par des cordes* (Paris, 1687), 1.

99. *The Works of Robert Boyle*, 1:249.

100. Boyle, 1:249.

101. Boyle, 1:249.

102. Boyle, 1:165.

103. Boyle, 1:249.

104. *Contemplation of Created Things*, 147.

6. "THROTTLED AT LENGTH IN THE AIR"

1. The volume appeared in print probably later that autumn but bears the publication date of 1671.

2. For the main political controversies of the year 1670, see Knoppers, "'Englands Case': Contexts of the 1671 Poems," 573–87, including a bill to allow Lord Roos's divorce (with seeming implications for the royal marriage and prevention of a Catholic heir) and an act to restrict conventicles, or the private meetings of nonconformists. Greaves, *Deliver Us from Evil*, identifies several separate groups that opposed the Restoration and mounted a decentralized resistance effort in a series of rebellion plots, and through other forms of dissent.

3. Weber, *Paper Bullets*, 153–61, describes a zealous campaign by Charles's chief regulator, Surveyor Roger L'Estrange, against unlicensed printing during 1660s and '70s, and the alternative forms of unregulated speech and print that surfaced in London coffee houses, leading to a royal crackdown on these spaces in 1675.

4. Milton's last stab at the genre of polemical political writing during the 1660s, *A Readie & Easie Way to Establish a Free Commonwealth*, was published in a second edition in April 1660, shortly before the accession of Charles II.

5. Rumrich and Fallon, "Introduction," 11.

6. On the metabolic process integral to the translation of matter from the grossly corporeal by degrees to a more refined and spirituous state, see Fallon, *Milton among the Philosophers*, 102–6.

7. For more on the morally pliant natural world, particularly the atmosphere, see chapter 1. For more on the subject of free choice and the physical world see, Fallon, *Milton among the Philosophers*, 198–202.

8. Hiltner, *What Else Is Pastoral?*, 97–101, offers a succinct description of the factors leading to London's air pollution crisis. The classic account of England's early turn to coal as fuel and, as the author argues, the resultant formation of capitalism, is Nef,

The Rise of the British Coal Industry. Cavert, *The Smoke of London*, 17–31, updates Nef's thesis to include subsequent research, and emphasizes that coal was central not only to economic expansion, but also to the development of English society.

9. In *Hubbub,* Cockayne documents an array of human-made nuisances in London and other English cities, including coal smoke and reeking dunghills, that provoked disgust, public outrage, and litigation.

10. Evelyn's *Fumifugium: of The Inconveniencie of the Aer and Smoak of London Dissipated* (1661), addressed directly to Charles II, was the first comprehensive analysis of London's air pollution problem. It treats the profuse coal smoke over the city not only as a nuisance, but also a cause of respiratory disease. Cavert, *The Smoke of London*, 175–82, adds the important caveat that Evelyn's proposals for reducing and containing smoke emissions, were primarily designed to protect royal interests, and hearkened back to pre-Civil War efforts to preserve the Carolinian court from symbolic degradation by smoke from London's breweries. On the medical grounds of Evelyn's argument, see Cavert, *The Smoke of London*, 86–87.

11. Hiltner stresses the role of repopulation following the Black Plague, deforestation, and inflated wood prices in driving the turn to coal; *What Else Is Pastoral?*, 97–98. Cavert qualifies the "timber crisis" thesis; *The Smoke of London*, 19–22.

12. Cavert, *The Smoke of London*, 24.

13. Cavert, *The Smoke of London*, 37. For a reconstruction of London's pollutant levels, see Peter Brimblecombe and Carlotta M. Grossi, "Millennium-long damage to building materials in London," *Science of the Total Environment* 407 (February 2009), 1354–61.

14. For a chronology and list of Milton's residences, see Dobranski, "'Where Men of Differing Judgements Croud,'" appendix A.

15. Fewer houses and westerly winds created better conditions in the West End; Cavert, *The Smoke of London*, 38. John Evelyn notes that winds brought pollution across the Thames into the City, and that more affluent areas to the west were spared the stink of the eastern part of the city; see Brimblecombe, *The Big Smoke*, 71–72.

16. On the remedies Bridgewater sought and documented in his correspondence during the summer of 1666, see Cavert, *The Smoke of London*, 61–63. On Milton's residence near Bridgewater House, see Denton, *Records of St. Giles' Cripplegate*, 159–60.

17. For examples of the means Londoners used to correct or receive compensation for the nuisance of coal smoke, see Cavert, *The Smoke of London*, 61–79. On nuisances more generally, see also, Cockayne, *Hubbub*, 18–21.

18. Masson, *The Life of John Milton*, 6:490, notes that a plague pit where bodies were thrown into a common grave opened up in Bunhill Fields steps from Milton's home in Artillery Walk. Milton left London before the number of plague deaths climbed above 20,000.

19. Hiltner, *What Else Is Pastoral?*, 104. Hiltner's reading of the description of hell in *Paradise Lost* (104–5) focuses on 1.350 and 1.670–74.

20. "Just as the essential ingredients of gunpowder . . . were the sulphur and that 'master-ingredient,' saltpeter or nitre, so these same substances reacting together in the clouds or beneath the surface of the earth could account for these familiar yet

terrifying natural phenomena"; Guerlac, "The Poets' Nitre," 246–47. I am indebted to John Heilbron for pointing out this reference to me.

21. Guerlac, 250–55.

22. Browne, *Pseudodoxia Epidemica*, 69.

23. Browne, 69.

24. "The Poets' Nitre," 250–51. Guerlac also points out that the "tumultuous cloud / Instinct with fire and nitre" (2.936–37) that Satan rides in chaos contains the meteorologically active pair of ingredients: sulfur ("fire") and "nitre" (247).

25. James I complained about the smoke damage to St. Paul's and Archbishop Laud fined brewers for the same; see Brimblecombe, "Air Pollution and Architecture," 35. Certain prominent chemists and physicians such as Sir Kenelm Digby and Gideon Harvey agreed with John Evelyn that the sulfurous sea coal burned in London contributed to respiratory disease; see Cavert, *The Smoke of London*, 86–87.

26. Jessey, *The exceeding riches of grace*, 77–78.

27. s.v. "brimstone," in vol. 1 of McClintock and Strong, eds., *Cyclopædia of Biblical, Theological, and Ecclesiastical Literature*. See also s.v. "sulphur," in Gesenius, Robinson, and Potter, *An English-Hebrew Lexicon* and s.v. "גָּפְרִית," in Feyerabend, *A complete Hebrew-English Pocket-Dictionary*. Hiltner argues that the *gophrith* of the story of the destruction of Sodom and Gomorrah at Genesis 19:24, "meant neither a stone nor coal nor sulfur, but rather 'Jehovah's breath'" and that the sulfurous connotations of such biblical passages were introduced with the English translation "brimstone," which, as he demonstrates, denotes "sulfur" or "burning stone" and can be used as a synonym for sea coal; *What Else Is Pastoral?*, 102–3. But several dictionaries, including *Brown-Driver-Briggs Hebrew and English Lexicon, Unabridged* and *Gesenius's Hebrew and Chaldee Lexicon to the Old Testament Scriptures*, suggest that the word *gophrith* was applied to sulfur from biblical times.

28. Jessey, *The exceeding riches of grace*, 78.

29. Johnstone, "The Protestant Devil," 179. Internal temptation by an invisible, spiritual devil rather than an obviously physical and "othered" adversary, Johnstone argues, was the focus of a Protestant brand of demonism in England.

30. Lawrence, *Of our communion and warre with angels*, 65.

31. Gouge, *The vvhole-armor of God*, 75, recognizes the conventional boundaries of Satan's place on earth: "the Divels dominion is restrained to the ayre." I quote the second edition.

32. Although Gouge allows Satan profound powers he is careful to follow Aquinas's rule (*Summa Theologiae*, part 1, question 110, article 4) that angels cannot themselves perform miracles nor act above the laws of created nature; *The vvhole-armor of God*, 70–71.

33. Gouge, 89–90.

34. See [Philaretes and Hind], *Work for Chimney-sweepers*, sigs. F4–G4v; in *Paradise Lost*, Milton portrays Satan as the inventor of gunpowder (6.470–506).

35. Gillis, *Still Turning*, 5. Ash, *The Draining of the Fens*, approaches the fenland drainage projects from the 1570s to the 1650s as a facet of state building carried out by projectors that encountered public opposition and technical setbacks along the way.

36. Jonson's play was first performed in 1616, the same year that *The vvhole-armor of God* was published.

37. *The Sky of our Manufacture*, 50.

38. For example, William Gurnall, *The Christian in compleat armour. Or, A treatise of the saints war against the Devil* [. . .] (London, 1655); Paul Baynes, *The spirituall armour With which being furnished, a Christian may be able to stand fast in the euill day, and time of tryall; and to quench all the fiery darts of the wicked* (London, 1620); Ralph Robinson, *Panoplia. Universa arma. Hieron. Or, The Christian compleatly armed* (London, 1656).

39. "Having Done All to Stand," 202.

40. The passage of *Paradise Regained* in question is 4.551–71 where "stand" occurs three times at 4.551 and once at 4.554; "stood" occurs twice at 4.561 and 4.571.

41. Brannon, *The Heavenlies in Ephesians*, 1.

42. Arnold, *Ephesians, Power and Magic*, 153–55.

43. Brannon, *The Heavenlies in Ephesians*, 198n40.

44. Lewalski, *Milton's Brief Epic*, 161.

45. See, for example, Hughes, "The Christ of 'Paradise Regained'," and Steadman, "Heroic Virtue and the Divine Image," which restricts its focus to book 7 of Aristotle's *Ethics*.

46. Augustine, *Books 8–11*, 345–47.

47. Augustine, 345; emphasis added.

48. Augustine, 345.

49. Augustine, *Books 8–11*, 347; see *Aeneid*, 3.438–39.

50. Augustine, *Books 8–11*, 347.

51. Speaking to his fellow devils, Satan cites "he surnamed of Africa" as an example of a man with a stronger-than-usual ability to resist temptation (2.199). Without committing to a judgment on Africanus, Jesus casts doubt on his military motives at 2.101–4.

52. *Paradise Lost*, 10.184–89. For the full quotation see chapter 3.

53. Gouge, *The vvhole-armor of God*, 89.

54. At 2.135–43, Satan expresses uncertainty as to whether Jesus is man or god, admitting that he seems to have divine gifts. At 4.197–203, Satan appears to have growing confidence that Jesus is "in higher sort / Than" men or angels. But he denies this at 4.536.

55. Theological commentators cited Satan's power as prince of the air to explain how he produced the vision of the kingdoms; see Partner, *Poetry and Vision*, 255–56; for relevant scientific contexts regarding the telescope and microscope in Milton's explanations, see Svendsen, "Milton's 'Aerie Microscope'" and Nicholson, "Milton and the Telescope," 11 (and note).

56. See Baldwin, "Note on *Il Penseroso*," 185.

57. Baldwin, "Note on *Il Penseroso*," 184–85. Finney, *Musical Backgrounds*, 103–4, suggests that the source of *Il Penseroso*'s idea that demons inhabit the elements is Ficino's commentary on Plato's *Symposium*.

58. *Hermetica*, 60.

59. "Elements" in Gal. 4:3 (KJV), quoted above, is the translation of στοιχεῖα (stoicheia), which means the first principles from which other parts proceed in order. *A Treatise of Civil Power* translates Gal. 4:3 using the word "rudiments" in place of KJV's

"elements." Tyndale uses "ordinaunces." Although *Of Civil Power* is mainly interested in how Galatians 4:3 argues that Christ nullifies certain ceremonial aspects of Mosaic law, στοιχεῖα has a non-legalistic reference in scripture, for example, at 2 Peter 3:10 and 3:12, where the word clearly refers to the basic physical elements of the world.

60. On the stoning incident from the Old Testament and for a contrasting reference to Elijah's encounter with a widow gathering sticks at 1 Kings 17:10, see Burnett, "Milton's 'Paradise Regained,' I.314–19." Carey, *Milton: The Complete Shorter Poems*, 436, note i 314, suggests John 15:6.

61. Hatcher, *The History of the British Coal Industry*, 31–32. Much of the following analysis of the timber crisis will be drawn from Hatcher's account.

62. Hatcher, 33–35.

63. Hatcher, 34.

64. Hatcher, 49.

65. Instead of deciding right away what to do with the man, the people take him into custody until Moses receives a ruling from on high: "And they put him in ward, because it was not declared what should be done to him" (Num. 15:34).

66. Governments used various instruments during the sixteenth and seventeenth centuries, including restrictive regulations and naval convoys, to keep the price of fuel stable. On the price fixing, fuel rationing, and laws aimed at keeping the price of wood at bay in Worcester, see Hatcher, *The British Coal Industry*, 48. On protection of the colliers, public morale, the volatility of coal prices, and predatory inflationary practices during the First and Second Dutch Wars, see Cavert, *The Smoke of London*, 163–66.

67. This was already the case in medieval London, although at that time the hinterlands were well wooded; Galloway, Keene, and Murphy, "Fuelling the City," especially 460–65.

68. *OED Online*, s.v. "stubble, n. 1," accessed February 15, 2020.

69. "Stubwood" is "all wood which grows in hedgerows and does not come under the denomination of 'timbers,' 'pollards,' or 'thorns'"; Marshall, *The Rural Economy of Norfolk*, 2:389. On the importance of hedgerows from a fuel gathering standpoint, see also Warde and Williamson, "Fuel supply and agriculture in post-medieval England," especially 67–69.

70. Taylor, "The Marriage Ring," 217.

71. In an organic economy, not only was land the sole "source of food, it was also the source directly or indirectly of all the material products of use to man"; Wrigley, *Energy and the English*, 9.

72. Organic economies are limited by the "fixed supply of land" or, another way of putting it, the "energy constraint," which is determined by the amount of energy produced from plant photosynthesis and captured by humans. Growth in production requiring thermal energy necessitated additional woodlands and thus cut into the area of land dedicated to growing food. See Wrigley, *Energy and the English*, 10–17, especially 10.

73. Wrigley, 28–42.

74. They complained that the poor would "starve" if price inflation did not abate and conflated scarcity of fuel could with a "dearth" of food; see Cavert, *The Smoke of London*, 111–13.

75. Cavert, 111.

76. This observation was made by Aeneas Sylvius, latterly known as Pope Pius II: "Nam pauperes pene nudos ad templa mendicantes, acceptis lapidibus eleemosynae gratia datis, laetos abiisse conspeximus: id genus lapidis sive sulphurea sive alia pingui materia praeditum, pro ligno, quo regio nuda est, com- buritur," *Aeneae Sylvii Piccolominei Senensis* [. . .] *Opera que extant omnia* [. . .], (Basel, 1551), 443, qtd. in Turner, "English Coal Industry," 2n9; see also Brand, *The History and Antiquities*, 2:263.

77. The raw product traveled by cart from the pithead to a wharf on the Tyne, then by sea to London where it traveled up the Thames. For a contemporary recipe for making coal briquettes, see Plat, *A new, cheape and delicate fire*.

78. The provision of coal to London's poor weighed heavily on civic and Parliamentary leaders throughout the early modern period as they feared fuel scarcity could lead to instability and even rebellion. See Cavert, *The Smoke of London*, 115–19.

79. Gardiner, *Englands grievance discovered*, 201.

80. On the history of the Hostmen who dominated the Newcastle coal trade, see Nef, *The Rise of the British Coal Industry*, 2:119–34. See also Newton, *North-East England*, 37–38.

81. Gardiner, *Englands grievance discovered*, 203.

82. As locals called the Hostmen who were the principal coal traders in the area; see Nef, *The Rise of the British Coal Industry*, 2:120.

83. "The *Owners of Colleries*, must first sell the Coals to the Magistrates of *Newcastle*, the *Magistrates* to the *Masters* of ships, the *Master* of *ships* to the *Woodmongers* or *Wharfingers*, and they to those that spend them: Every change of the propriety adding to, and enhancing the price of the Coals, thus interchangeably bought and sold; which course, as it picks some money out of the purses of every man that buys Coals, besides bad Coals being therby vented"; Gardiner, *Englands grievance discovered*, 201.

84. In *Paradise Lost* Milton invokes Arabian spices and their effect on those sailing along trade routes to describe Satan's initial captivation with the sweet-smelling air of Eden (4.163).

85. Emphasis added.

86. Thomas Adams, *The deuills banket described in foure sermons* (London, 1614), 7.

87. Aulus Vitellius, 15–69 CE, was briefly emperor of Rome in 69, suffering a humiliating defeat the same year. Tacitus notes the emperor's gluttony and luxury, and the lust and cruelty of his followers; *The Histories*, 2.275–77.

88. Apicius, for example. See Juvenal, "Satire 11."

89. Giuseppe Arcimboldo (ca. 1527–1593), an Italian portraitist in the courts of three Holy Roman Emperors, known for painting his subjects with heads composed of vegetables, fruits, flowers, and fish.

90. The estimate is Nicolas Monardes's, a sixteenth-century Spanish physician, paraphrased in Dannenfeldt, "Ambergris: The Search for Its Origin," 388. According to Dannenfeldt, famous travel literature such as Gian Battista Ramusio's mid-sixteenth-century edition of Marco Polo, describes whaling in coastal regions, in this case the Arabian Sea, to extract oil and ambergris (386).

91. The Romans sometimes gave hyperbolic names to favorite dishes, e.g., *cere-

brum Jovis or *clypeum Minervae;* see William Warburton's note (line 346) in Todd, ed., *The Poetical Works of John Milton,* 5.121.

92. "Mullus erit domini quem misit Corisca vel quem / Tauromenitanae rupes, quando omne peractum est / et iam defecit nostrum mare, dum gula saevit, / retibus adsiduis penitus scrutante macello / proxima, nec patimur Tyrrhenum crescere piscem" (92–95); "Satire 5," 222.

93. *OED Online,* s.v. "tender, v.1, 1.a.," accessed February 27, 2020: to offer formally, specifically, to offer (e.g., money) in order to discharge a debt.

94. Cp. "toil'st in vain" (4.498). "Toils" has a double meaning: a hunter's net ("toil, n.2, 1.b."), but also "something produced or accomplished by hard or prolonged exertion"; s.v. "toil, n.1, 2.b.," *OED Online,* accessed March 1, 2020. See also "Come forth, And tast the ayre of Palaces, eate, drinke The toyles of Empricks, and their boasted practise: Tincture of Pearle, and Corall, Gold, and Amber"; Jonson, *The Alchemist,* 4.1., sig. I.

95. The river is the now called the Kerkha. Athenaeus, *Deipnosophists,* book 12, qtd. in Carey, *Milton: The Complete Shorter Poems,* 475n iii288–89.

96. Milton mentions Cyrus just four lines earlier at (4.284). Herodotus, *The Histories,* 1.188.2, qtd. in Carey, *Milton: The Complete Shorter Poems,* 475n iii288–89.

97. The scarcity of fertile land in fact compelled the ancient Greeks to explore and expand their territory in search of richer tillage. Elvio Cifferi, s.v. "Greek Empire," in *Encyclopedia of World Geography,* 1:33.

98. Quint, "The Disenchanted World," 183.

99. Quint, 184.

100. Quint, 186–87, 190.

101. Quint, 192.

102. See also the lines, "Satan bowing low / His grey dissimulation, disappeared / Into thin air diffused" (497–99), where instead of using a vague term like *vanish,* Milton specifies Satan's color and the physical process whereby he takes leave of Christ at the end of book 1.

103. By emphasizing the continuity rather than the discontinuity between the storm episode and previous trials, I differ with Steadman, "'Like Turbulencies,'" 86, who argues that the storm episode represents "an altogether different type of temptation" from the previous examples, because instead of promising Jesus prosperity, it subjects him to adversity.

104. For instance, Palma Vecchio's painting, discussed in chapter 3.

105. An improbable pair, both Milton and the figure of Sin experience these circumstances in *Paradise Lost.* At 2.801, Sin's howling hound-offspring "with conscious terrors vex [her] round," and at 7.27, Milton portrays himself "with dangers compassed round" such as evil speech and the dissonant sounds of the rabble.

106. "This tempest at this desert most was bent / Of men at thee, for only thou here dwell'st" (4.465–66).

107. According to Quint, "The Disenchanted World," 192, Jesus does not acknowledge "any metaphysical content" in the storm. His subsequent triumph on the pinnacle dispels the false human notion that Nature was ever the possession of demons

and is proof that "[t]he romance of the world is over, true science taking the place of 'the science of magic'" (193).

108. "Having Done All to Stand," 203.

109. However, previous glosses on the simile have tended to focus on a different dynamic—the traditional Renaissance interpretation that the myth represents the triumph of the spirit or reason, personified by the Hercules / Christ figure, over the flesh, embodied by Antaeus / Satan. For example, Lewalski, *Milton's Brief Epic*, 319, builds on the traditional interpretation to suggest that the simile symbolizes a victor who eschews the need for worldly things, expels devils, and gives up his own mortal life; Swaim, "Hercules, Antaeus, and Prometheus," also emphasizes the reason-conquers-worldliness motif adding analysis of Hercules's journey to free Prometheus after he kills Antaeus, which she argues refers to the new dispensation accomplished by Christ's passion (151–52).

110. "The Typology of 'Paradise Regained'," 237.

111. "Hercules, Antaeus, and Prometheus," 143.

112. On other classical sources of the myth and later commentaries, see Brumble, *Classical Myths and Legends*, 26–27.

113. "Rapit arida tellus / Sudorem: calido conplentur sanguine venae, / Intumuere tori, totosque induruit artus / Herculeosque novo laxavit corpore nodos"; Lucan, *The Civil War*, 4.220–21.

114. "Auxilium membris calidas infudit harenas"; Lucan, 4.220–21.

115. "in terrae autem reliquis generibus vel maxime mira natura est sulpuris, quo plurima domantur"; *Natural History*, 9:388–89.

116. Pliny, 9:391–93.

117. "Quisquis inest terris in fessos spiritus artus / Egeritur, Tellusque viro luctante laborat"; Lucan, *The Civil War*, 4.222–23.

118. *Natural History*, 9:1–5. On the Romans' moral objections to gold mining, see Healy, "Pliny the Elder and Ancient Mineralogy," 174.

119. Evelyn, *Fumifugium*, 11.

120. Ovid, *Metamorphoses*, 9.183–84; Apollodorus, *The Library*, 2.5.11; Lucan, *The Civil War*, 4.652–53.

121. Comes, *Mythologiae*, 7.1. qtd. in Starnes and Talbert, *Classical Myth And Legend*, 238; Fraunce, *The Countess of Pembrokes Yvychurch*, 46v, qtd. in Brumble, *Classical Myths and Legends*, 26–27.

122. "Scruzd," meaning squeezed, is the method Arthur uses to kill the Antaeus figure Maleger in the *The Faerie Queene*, 2.11.36.

123. *OED Online*, 3rd ed., s.v., "throttle, v.," accessed March 16, 2020.

124. *OED Online*, 3rd ed., s.v., "throttle, n. I.," accessed March 16, 2020; Rowlands, *Greenes ghost haunting conie-catchers*, 9r.

125. Evelyn, *Fumifugium*, 16.

126. Evelyn, 10.

127. So it is called in [Oldenburg and Beale,] *"Observations continued."*

128. This term of art persists in barometry through the mid-nineteenth century; see *OED Online*, 3rd ed., s.v. "station, n., 4.b.," accessed March 19, 2020; Oldenburg to Boyle, October 27, 1664, in *The Correspondence of Robert Boyle*, 2:366–69.

129. Oldenburg to Boyle, *The Correspondence of Robert Boyle*, 2:366–67.

130. See "The Preface," in *Micrographia*. In *"A new Contrivance of* Wheel-Barometer," Hooke amends the description of his wheel barometer from the *Micrographia*, giving an easier way to improve "any *Common Baroscope*" (218).

131. For example, an extract of a letter by John Beale offers insight on the ascent and descent of the barometer: "Again, if the *Mercury* ascends to a good height after the fall of rain (as sometimes, but less often it does) then I look for a settled serenity; but if it proceeds after rain in a descending motion, then I expect a continuance of broken and showry weather"; *"Observations continued,"* 165.

132. [Oldenburg and Beale], *"Observations continued,"* 164.

Bibliography

WORKS BY MILTON

All quotations of works by Milton are from the following editions, except as noted.

The Complete Poetry and Essential Prose of John Milton. Edited by William Kerrigan, John Rumrich, and Stephen M. Fallon. New York: Modern Library, 2007.
Complete Prose Works of John Milton. Edited by Don M. Wolfe. 8 vols. New Haven, CT: Yale University Press, 1953–82.
De Doctrina Christiana, part 1. Edited by John K. Hale and J. Donald Cullington. Vol. 8 of *The Complete Works of John Milton,* edited by Thomas N. Corns and Gordon Campbell. Oxford: Oxford University Press, 2008–.
The Essential Prose of John Milton. Edited by William Kerrigan, John Rumrich, and Stephen M. Fallon. New York: Random House, 2007. Reprinted by Modern Library, New York, in 2013.
A Mask Presented at Ludlow-Castle, 1634. In *Poems of Mr. John Milton.* London, 1645.
Milton: The Complete Shorter Poems. 2nd ed. Edited by John Carey. Harlow, England: Longman, 2007.
Paradise Lost. 2nd ed. Edited by Alastair Fowler. London: Longman, 1998.
Paradise Lost: A Poem in Twelve Books. London, 1674.
Second Defence of the People of England. Edited by Eugene J. Strittmatter and translated by George Burnett. Vol. 8 of *The Works of John Milton,* edited by Frank Allen Patterson. New York: Columbia University Press, 1933.

REFERENCES

Adams, Thomas. *The deuills banket described in foure sermons.* London, 1614.
Agrippa, Heinrich Cornelius. *Of the Vanitie and Uncertaintie of Artes and Sciences.* Translated by James Sanford. London, 1569.
———. *Three Books of Occult Philosophy, written by Henry Cornelius Agrippa, of Nettesheim [. . .].* Translated by J. F. [John French]. London, 1651.
Algra, Keimpe. "Stoics on Souls and Demons: Reconstructing Stoic Demonology." In *Demons and the Devil in Ancient and Medieval Christianity,* 71–96, edited by Nienke Vos and Willemien Otten. Leiden, The Netherlands: Brill, 2011.

Algra, Keimpe, Jonathan Barnes, Jaap Mansfeld, and Malcolm Schofield, eds. *The Cambridge History of Hellenistic Philosophy*. Cambridge: Cambridge University Press, 1999.

Anderson, Katharine. *Predicting the Weather: Victorians and the Science of Meteorology*. Chicago: University of Chicago Press, 2010.

Antinori, Vincenzo. *Notizie istoriche relative all'Accademia del Cimento*, in the series Saggi di Naturali esperienze fatte nell'Accademia del Cimento. Florence, 1841.

Aquinas, Thomas. "Expositio super Job ad litteram." In *S. Thomae Aquinatis Opera Omnia: ut sunt in indice thomistico, additis 61 scriptis ex aliis medii aevi auctoribus*, vol. 5, *Commentaria in scripturas*, 1–49, edited by Roberto Busa. 7 vols. Stuttgart-Bad Cannstatt: Frommann-Holzboog, 1980.

———. *The Literal Exposition of Job: A Scriptural Commentary Concerning Providence*. Translated by Anthony Damico. Vol. 38 of *The Collected Works of St. Thomas Aquinas*. Electronic ed. 47 vols. Charlottesville, VA: InteLex Corp., 1993. Past Masters online database.

Aristotle, *De Generatione et Corruptione*. Translated by C. J. F. Williams. Oxford: Clarendon Press, 1982.

———. *Meteorologica*. Translated by H. D. P. Lee. Cambridge, MA: Harvard University Press, 1952.

———. *On the Heavens*. Translated by W. K. C. Guthrie. Cambridge, MA: Harvard University Press, 1939.

Arnold, Clinton E. *Ephesians: Power and Magic: The Concept of Power in Ephesians in Light of its Historical Setting*. Cambridge: Cambridge University Press, 1989.

Ash, Eric H. *The Draining of the Fens: Projectors, Popular Politics and State Building in Early Modern England*. Johns Hopkins Studies in the History of Technology. Baltimore: Johns Hopkins University Press, 2017.

Athanasius. "The Life and Affairs of Our Holy Father Antony." In *Athanasius: The Life of Antony and The Letter to Marcellinus*, 29–100, edited and translated by Robert C. Gregg. New York: Paulist Press, 1980.

Augustine. *Books 8–11*. Translated by David S. Wiesen. Vol. 3 of *The City of God against the Pagans*, edited by George E. McCracken. 7 vols. London: William Heinemann, 1968.

———. *Books 1–6*. Vol. 1 of *St. Augustine: The Literal Meaning of Genesis*, edited and translated by John Hammond Taylor. 2 vols. New York: Newman Press, 1982.

Bacon, Francis. *The Works of Francis Bacon*. Edited by James Spedding, Robert Ellis, and Douglas Heath. 14 vols. London: Longman, 1857–74.

———. *Of the Dignity and Advancement of Learning*. In vol. 4 of *The Works of Francis Bacon*, edited by James Spedding, Robert Ellis, and Douglas Heath, 275–498. London: Longman, 1857–74.

———. *Sylva Sylvarum*. In vol. 2 of *The Works of Francis Bacon*, edited by James Spedding, Robert Ellis, and Douglas Heath, 339–672. London: Longman, 1857–74.

Baldwin, Edward Chauncey. "A note on *Il Penseroso*." *Modern Language Notes* 33, no. 3 (1918): 184–85.

Barnett, Pamela R. *Theodore Haak, F.R.S. (1605–1690): The First German Translator of Paradise Lost*. The Hague, The Netherlands: Mouton & Co., 1962.

Behringer, Wolfgang. "Climatic Change and Witch-Hunting: The Impact of the Little Ice-Age on Mentalities." *Climatic Change* 43, no. 1 (September 1999): 335–51.

———. *A Cultural History of Climate Change.* Translated by Patrick Camiller. Cambridge: Polity Press, 2010.

Beeckman, Isaac. *Journal tenu par Isaac Beeckman de 1604 à 1634 publié avec un introduction et des notes par C. De Waard.* Edited by C. de Waard. 4 vols. The Hague, The Netherlands: Martinus Nijhoff, 1939–53.

Berley, Marc. *After the Heavenly Tune: English Poetry and the Aspiration to Song.* Pittsburgh, PA: Duquesne University Press, 2000.

Blair, Ann. *The Theater of Nature: Jean Bodin and Renaissance Science.* Princeton: Princeton University Press, 1997.

Boas, Marie. "The Establishment of the Mechanical Philosophy." *Osiris* 10 (1952): 412–541.

Bodin, Jean. *Universae Naturae Theatrum.* Lyon, [France,] 1596.

Boyle, Robert. *The Correspondence of Robert Boyle, Volume 1, 1636–61, Introduction.* Edited by Michael Hunter, Antonio Clericuzio, and Lawrence M. Principe. London: Pickering & Chatto, 2001.

———. *The Correspondence of Robert Boyle,* edited by Michael Hunter, Antonio Clericuzio, and Lawrence Principe. Vol. 2: *1662–5.* London: Routledge, 2016. First published in 2001 by Pickering & Chatto.

———. "*Of a* New *kind of* Baroscope, *which may be call'd* Statical [. . .]." *Philosophical Transactions* 1, no. 14 (July 2, 1666): 231–39.

———. "Some Observations and Directions about the Barometer, communicated by the same Hand, to the Author of this Tract." *Philosophical Transactions* 11 (April 2, 1666): 181–85.

———. *The Works of Robert Boyle.* Edited by Michael Hunter and Edward B. Davis. 14 vols. London: Pickering & Chatto, 1999.

Brand, John. *The History and Antiquities of the Town and County of the Town of Newcastle upon Tyne.* 2 vols. London, 1789.

Brannon, M. Jeff. *The Heavenlies in Ephesians: A Lexical, Exegetical, and Conceptual Analysis.* Library of New Testament Studies. New York: T&T Clark International, 2011.

Breasted, Barbara. "'Comus' and the Castlehaven Scandal." *Milton Studies* 3 (1971): 201–24.

Brimblecombe, Peter. "Air Pollution and Architecture: Past, Present and Future." *Journal of Architectural Conservation* 6, no. 2 (July 2000): 30–46.

———. *The Big Smoke: A History of Air Pollution in London since Medieval Times.* Routledge Revival Edition. Abingdon, UK: Taylor and Francis, 2011.

Brumble, H. David. *Classical Myths and Legends in the Middle Ages and Renaissance: A Dictionary of Allegorical Meanings.* London: Routledge, 1998.

Brown, Cedric C. *John Milton's Aristocratic Entertainments.* Cambridge: Cambridge University Press, 1985.

———. "Milton's Ludlow Masque." In *The Cambridge Companion to Milton,* edited by Dennis Danielson, 25–38. Cambridge: Cambridge University Press, 1999.

Browne, Thomas. *Pseudodoxia Epidemica: Or, Enquiries Into Very Many Received Tenents* [. . .]. 3rd ed. London, 1658.

Buhler, Stephen M. "Counterpoint and Controversy: Milton and the Critiques of Polyphonic Music." *Milton Studies* 36 (1998): 18–40.

Burnett, Archie. "Milton's 'Paradise Regained', I.314–19." *Notes & Queries* 25, no. 6 (December 1978): 509–10.

Burnett, Charles, Michael Fend, and Penelope Gouk, eds. *The Second Sense: Studies in Hearing and Musical Judgement from Antiquity to the Seventeenth Century*. London: The Warburg Institute University of London, 1991.

Burnett, Charles. "Sound and its Perception in the Middle Ages." In *The Second Sense*, edited by Burnett, Fend, and Gouk, 43–69.

Burton, Robert. *The Anatomy of Melancholy. What it is, with all the kinds causes, symptomes, [. . .] of it.* 5th ed. Oxford, 1638.

Bush, Douglas Earl, and Richard Kassel. *The Organ: An Encyclopedia*. New York: Routledge, 2006.

Butler, Charles. *The Principles of Musik, in Singing and Setting: With the Two-fold use Thereof.* London, 1636.

Butler, George F. "Milton's Meeting with Galileo: A Reconsideration." *Milton Quarterly* 39, no. 3 (October 2005): 132–39.

Cajori, Florian. *A History of Physics in its Elementary Branches, including the Evolution of Physical Laboratories*. New York: MacMillan, 1924.

Campbell, Murray, Clive Greated, and Arnold Myers, eds. *Musical Instruments: History, Technology, and Performance of Instruments of Western Music*. Oxford: Oxford University Press, 2004.

Camuffo, Dario, Chiara Bertolin, Phil D. Jones, Richard Cornes, and Emmanuel Garnier. "The earliest daily barometric pressure readings in Italy: Pisa AD 1657–1658 and Modena AD 1694, and the weather over Europe." *Holocene* 20, no. 3 (2010): 337–49.

Cavert, William M. *The Smoke of London: Energy and Environment in the Early Modern City*. Cambridge: Cambridge University Press, 2016.

Charleton, Walter. *Physiologia Epicuro-Gassendo-Charltoniana, or, A fabrick of science natural, upon the hypothesis of atoms founded by Epicurus [. . .]*. London, 1654.

Cockayne, Emily. *Hubbub: Filth, Noise, & Stench in England, 1600–1770*. New Haven, CT: Yale University Press, 2008.

Cohen, Bernard. *The Newtonian Revolution*. Cambridge: Cambridge University Press, 1980.

Cohen, H. F. *Quantifying Music: The Science of Music at the First Stage of the Scientific Revolution, 1580–1650*. Dordrecht: D. Reidel Publishing, 1984.

Coleman, Satis N. *Bells: Their History, Legends, Making, and Uses*. Chicago: Rand McNally & Company, 1928.

Costello, William. *The Scholastic Curriculum at Early Seventeenth-Century Cambridge*. Cambridge, MA: Harvard University Press, 1958.

Cox, J. Charles. *Cornwall*. London: G. Allen & Company, Ltd., 1912.

Dales, Richard C. *The Scientific Achievement of the Middle Ages*. Philadelphia: University of Pennsylvania Press, 1973.

Daniel, Samuel. "Sonnet XLIII." In *Delia*. London: 1592.

Danielson, Dennis. "Astronomy." In *Milton in Context*, edited by Stephen Dobranski, 213–25. Cambridge: Cambridge University Press, 2010.

Dannenfeldt, Karl H. "Ambergris: The Search for Its Origin." *Isis* 73, no. 3 (September 1982): 382–97.
[Dati, Carlo] Smarrito, *Lettera a Filaleti di Timauro Antiate, Della vera storia della cicloide, e della famosissima esperienza dell'argento viuo.* Firenze: All' Insegna della Stella, 1663.
Denton, W. *Records of St. Giles' Cripplegate.* London: George Bell & Sons, 1883.
Desaguliers, John Theophilus. *A Course of Experimental Philosophy.* 2 vols. London, 1734–44.
Descartes, René. *Principles of Philosophy.* Translated by Valentine Rodger Miller and Reese P. Miller. Dordrecht: Kluwer Academic Publishers, 1982.
Dickson, Arthur. *Valentine and Orson: A Study in Late Medieval Romance.* New York: Columbia University Press, 1929.
Dobranski, Stephen B. "'Where Men of Differing Judgements Croud': Milton and the Culture of the Coffee Houses." *Seventeenth Century* 9, no. 1 (1994): 35–56.
Donne, John. "A Valediction: forbidding Mourning." In *John Donne: The Complete English Poems*, 84. London: Penguin, 1996
Duran, Angelica. *The Age of Milton and the Scientific Revolution.* Pittsburgh: Duquesne University Press, 2007.
Edwards, Karen. *Milton and the Natural World: Science and Poetry in "Paradise Lost."* Cambridge: Cambridge University Press, 1999.
Eichholz, D. E. "Aristotle's Theory of the Formation of Metals and Minerals." *Classical Quarterly* 43 (1949): 141–46.
Eliot, T. S. "A Note on the Verse of John Milton." In *Essays and Studies by Members of The English Association* 21, 32–40. Oxford: Clarendon Press, 1936.
———. "Milton I." In *On Poetry and Poets*, 156–64. 7th Noonday paperbound edition. New York: Farrar, Straus & Giroux, 1969. First printed in London: Faber and Faber, 1957.
Elledge, Scott, ed. *Paradise Lost. An Authoritative Text, Backgrounds and Sources, Criticism.* By John Milton. 2nd ed. Norton: New York, 1993.
Evans, J. Martin. *The Miltonic Moment.* Lexington: University Press of Kentucky, 1998.
Evelyn, John. *The Diary of John Evelyn.* Edited by William Bray. Washington, DC: M. Walter Dunne, 1901.
———. *Fumifugium: or The Inconveniencie of the Aer and Smoak of London Dissipated.* London, 1661.
Fallon, Stephen M. *Milton among the Philosophers: Poetry and Materialism in Seventeenth-Century England.* Ithaca, NY: Cornell University Press, 1991.
Feyerabend, Karl. *A complete Hebrew-English Pocket-Dictionary to the Old Testament.* Berlin-Schöneberg: G. Langenscheidt, [1905].
Ficino, Marsilio. *Opera Omnia.* Basel, 1576.
———. *Three Books on Life.* Translated by Carol V. Kaske and John R. Clark. Binghamton, NY: Medieval & Renaissance Texts & Studies in conjunction with The Renaissance Society of America, 1989.
Fink, Z. S. "Milton and the Theory of Climatic Influence." *Modern Language Quarterly* 2, no. 1 (March 1941): 67–80.
Finney, Gretchen Ludke. *Musical Backgrounds for English Literature: 1580–1650.* New Brunswick, NJ: Rutgers University Press, 1962.

Fitzgerald, Allan D., ed. *Augustine through the Ages: An Encyclopedia*. Grand Rapids, MI: William B. Eerdmans Publishing, 1999.

Flannagan, Roy, ed. *The Riverside Milton*. Boston: Houghton Mifflin Company, 1998.

Fleming, James Rodger. *Historical Perspectives on Climate Change*. Oxford: Oxford University Press, 2005.

Forsyth, Neil. "Having Done All to Stand: Biblical and Classical Allusion in *Paradise Regained*." Milton Studies 21 (1985): 199–214.

Frisinger, H. Howard. *The History of Meteorology to 1800*. New York: Neal Watson Academic, 1977.

Froidmondi, Liberti. *Meteorologicorum Libri Sex*. London, 1656.

Frye, Northrop. "The Typology of 'Paradise Regained'." *Modern Philology* 53, no. 4 (May 1956): 227–38.

Frytsche, Marcus. *Meteorum, hoc est Impressionum Aerearum et Mirabilium Naturae Operum*. Wittenberg: Cratoniana, 1598.

Fulke, William. *A goodly gallerye with a most pleasaunt prospect, into the garden of naturall contemplation, to behold the naturall causes of all kynde of meteors* [. . .]. London, 1563.

Furley, David. "Cosmology." In *Cambridge History of Hellenistic Philosophy*, edited by Algra, Barnes, Mansfeld, and Schofield, 412–51.

Galileo. *Le Opere di Galileo Galilei*. Edizione Nazionale. 20 vols. Florence, 1890–1909.

———. *Two New Sciences: Including Centers of Gravity & Force of Percussion*. Translated by Stillman Drake. Madison, WI: University of Wisconsin Press, 1974.

Galloway, James A., Derek Keene, and Margaret Murphy. "Fuelling the City: Production and Distribution of Firewood and Fuel in London's Region, 1290–1400." *Economic History Review* 49, no. 3 (August 1996): 447–72.

Garber, Daniel. "Remarks on the Pre-History of the Mechanical Philosophy." In *The Mechanization of Natural Philosophy*, edited by Roux and Garber, 3–26.

Gardiner, Ralph. *Englands grievance discovered, in relation to the coal-trade* [. . .]. London, 1655.

Gattei, Stephano. "Galileo's legacy: a critical edition and translation of the manuscript of Vincenzo Viviani's *Grati Animi Monumenta*." *British Journal for the History of Science* 50, no. 2 (2017): 181–228.

Gesenius, Wilhelm, Edward Robinson, and Joseph Lewis Potter. *An English-Hebrew Lexicon, being a Complete Verbal Index to Gesenius' Hebrew Lexicon*. New York: Hurd and Houghton, 1877.

Giles, Brian D. "Pressure, Surface." In *Encyclopedia of World Climatology*, edited by J. E. Oliver, 583–88. Dordrecht: Springer, 2005.

Gillis, Christopher C. *Still Turning: A History of Aermotor Windmills*. College Station: Texas A&M University Press, 2015.

Gliozzi, Mario. "Torricelli, Evangelista." In vol. 13 of *Complete Dictionary of Scientific Biography*, edited by Charles Coulston Gillispie, 433–40. 27 vols. Detroit, MI: Charles Scribner's Sons, 2008.

Golinski, Jan. *British Weather and the Climate of Enlightenment*. Chicago: Chicago University Press, 2007.

Gouge, William. *The vvhole-armor of God: or A Christians spiritual furniture, to keepe him safe from all the assaults of Satan* [. . .]. London, 1619.

Gouk, Penelope. "Music in Francis Bacon's Natural Philosophy." In *Number to Sound*, edited by Gozza, 135–52.
———. "Music, Melancholy, and Medical Spirits in Early Modern Thought." In *Music as Medicine: The History of Music Therapy since Antiquity*, edited by Peregrine Horden, 173–94. Aldershot, UK: Ashgate, 2000.
———. *Music, Science and Natural Magic in Seventeenth-Century England*. New Haven: Yale University Press, 1999.
———. "Some English Theories of Hearing in the Seventeenth Century: Before and After Descartes." In *The Second Sense*, edited by Burnett, Fend, and Gouk, 95–113.
———. "Transforming Matter, Refining the Spirit: Alchemy, Music and Experimental Philosophy around 1600." *European Review* 21, no. 2 (2013): 147–49.
Gozza, Paolo, ed. *Number to Sound: The Musical Way to the Scientific Revolution*. Dordrecht: Kluwer Academic Publishers, 2000.
Grant, Edward. *A History of Natural Philosophy: From the Ancient World to the Nineteenth Century*. Cambridge: Cambridge University Press, 2007.
Greaves, Robert L. *Deliver Us from Evil: The Radical Underground in Britain, 1660–1663*. Oxford: Oxford University Press, 1986.
Guerlac, Henry. "The Poets' Nitre." *Isis* 45, no. 3 (September 1954): 243–55.
Hainsworth, J. D. "Ups and Downs in *Paradise Lost*." *Essays in Criticism* 33, no. 2 (April 1983): 99–107.
Harding, David P. *The Club of Hercules: Studies in the Classical Background of* Paradise Lost, Illinois Studies in Language and Literature, vol. 50. Urbana: University of Illinois, 1962.
Hatcher, John. *The History of the British Coal Industry*, vol. 1: *Before 1700: Towards the Age of Coal*. Oxford: Oxford University Press, 1993.
Healy, John F. "Pliny the Elder and Ancient Mineralogy." *Interdisciplinary Science Reviews* 6, no. 2 (1981): 166–80.
Heilbron, J. L. *Galileo*. Oxford: Oxford University Press, 2010.
Heninger, S. K. *A Handbook of Renaissance Meteorology with Particular Reference to Elizabethan and Jacobean Literature*. Durham, NC: Duke University Press, 1960.
Henry, John. *The Scientific Revolution and the Origins of Modern Science*. 2nd ed. Houndmills, UK: Palgrave, 2002.
Hequembourg, Stephen. "Milton's 'Unoriginal' Voice: Quotation Marks in *Paradise Lost*." *Modern Philology* 112, no. 1 (2014): 154–78.
Hermetica: The Greek Corpus Hermeticum *and the Latin* Asclepius *in a New English Translation, with Notes and Introduction*. Edited by Brian P. Copenhaver. Cambridge: Cambridge University Press, 1992.
Hiltner, Ken. *Milton and Ecology*. Cambridge: Cambridge University Press, 2003.
———, ed. *Renaissance Ecology: Imagining Eden in Milton's England*. Pittsburgh: Duquesne University Press, 2008.
———. *What Else Is Pastoral? Renaissance Literature and the Environment*. Ithaca, NY: Cornell University Press, 2011.
Hollander, John. *The Untuning of the Sky: Ideas of Music in English Poetry, 1500–1700*. Princeton, NJ: Princeton University Press, 1961.

Hooke, Robert. *Micrographia, Or, Some Physiological Descriptions of Minute Bodies Made by Magnifying Glasses: With Observations and Inquiries Thereupon*. London, 1665.

———. "A new Contrivance of Wheel-Barometer, much more easy to be prepared, than that, which is described in the Micrography; imparted by the author of that book." *Philosophical Transactions* 1, no. 13 (1665–66), 218–19.

Hooykaas, R. "Beeckman, Isaac." In *Complete Dictionary of Scientific Biography*, 1:566–68.

Hoxby, Blair. "The Wisdom of Their Feet: Meaningful Dance in Milton and the Stuart Masque." *English Literary Renaissance* 37 (2007): 74–99.

Hughes, Merritt Y. "The Christ of 'Paradise Regained' and the Renaissance Heroic Tradition." *Studies in Philology* 35, no. 2 (April 1938): 254–77.

Hunt, Frederick Vinton. *Origins in Acoustics: The Science of Sound from Antiquity to the Age of Newton*. New Haven: Yale University, 1978.

Hunter, William B. "Eve's Demonic Dream." *English Literary History* 13 (1946): 255–65.

Hutton, James. "Some English Poems in Praise of Music." *English Miscellany* 2 (1951): 1–63. Reprinted in *Music and the Renaissance: Renaissance, Reformation and Counter-Reformation*, edited by Philippe Vendrix, 145–208. Surrey, UK: Ashgate, 2011.

Jancović, Vladimir. *Reading the Skies: A Cultural History of English Weather, 1650–1820*. Chicago: University of Chicago Press, 2000.

Jessey, Henry. *The exceeding riches of grace advanced by the spirit of grace, in an empty nothing creature viz. Mrs. Sarah Wight, lately hopeles and restles, her soule dwelling as far from peace or hopes of mercy, as ever was any [. . .]*. London, 1647.

Johnstone, Nathan. *The Devil and Demonism in Early Modern England*. Cambridge: Cambridge University Press, 2006.

———. "The Protestant Devil: The Experience of Temptation in Early Modern England." *Journal of British Studies* 43, no. 2 (April 2004): 173–205.

Jonson, Ben. *The Alchemist*. London, 1612.

Juvenal, "Satire 5." In *Juvenal and Persius*, edited and translated by Susanna Morton Braund, 213–29. Cambridge, Mass.: Harvard University Library, 2004.

Kelley, Maurice, and Samuel D. Atkins. "Milton's Annotations of Aratus." *PMLA* 70 (December 1955): 1090–1106.

Kerrigan, William, John Rumrich, and Stephen M. Fallon, eds. *The Complete Poetry and Essential Prose of John Milton*. New York: Modern Library, 2007.

Kerrigan, William. *The Sacred Complex: On the Psychogenesis of "Paradise Lost."* Cambridge, MA: Harvard University Press, 1983.

Kingsley, Peter. *Ancient Philosophy, Mystery, and Magic: Empedocles and Pythagorean Tradition*. Oxford: Clarendon Press, 1995.

Kircher, Athanasius. *Musurgia Universalis sive Ars Magna Consoni et Dissoni in X Libros Digesta [. . .]*. 2 vols. Rome, 1650.

Klein, Robert. "Spirito Peregrino." In *Form and Meaning: Essays on the Renaissance and Modern Art*, translated by Madeline Jay and Leon Wieseltier, 62–85. New York: The Viking Press, 1970.

Knoppers, Laura Lunger. *The Oxford Handbook of Milton*, edited by Nicholas McDowell and Nigel Smith, 571–88. Oxford: Oxford University Press, 2011.

Kuhn, Thomas. "Reflections on My Critics." In *The Road Since Structure: Philosophical Essays, 1970–1993, with an Autobiographical Interview*, edited by Jim Conant and John Haugeland, 123–75. Chicago: University of Chicago Press, 2000.

Kupperman, Karen. "Climate and Mastery of the Wilderness in Seventeenth-Century New England." In *Seventeenth-Century New England*, edited by David D. Hall, David Grayson Allen, and Philip Chadwick Foster Smith, 3–37. Publications of the Colonial Society of Massachusetts, vol. 63. Boston: The Colonial Society of Massachusetts, 1984. https://www.colonialsociety.org/node/1742#ch01.

Latour, Bruno. *We Have Never Been Modern*. Translated by Catherine Porter. Cambridge, MA: Harvard University Press, 1993.

Lattis, James M. *Between Copernicus and Galileo: Christoph Clavius and the Collapse of Ptolemaic Cosmology*. Chicago: University of Chicago Press, 1994.

Lawrence, Henry. *Of our communion and warre with angels being, certaine meditations on that subject, bottom'd particularly (though not concluded within the compass of that Scripture) on Ephes. 6. 12 with the following verses, to the 19th*. [London], 1646.

Lawrence-Mathers, Anne. *Medieval Meteorology: Forecasting the Weather from Aristotle to the Almanac*. Cambridge: Cambridge University Press, 2019.

Leavis, F. R. "Milton's Verse." *Scrutiny* 2 (1933): 123–36.

Lenton, Timothy M., Hermann Held, Elmar Kriegler, Jim W. Hall, Wolfgang Lucht, Stefan Rahmstorf, and Hans Joachim Schellnhuber. "Tipping elements in the Earth's climate system." *Proceedings of the National Academy of Sciences* 105, no. 6 (February 12, 2008): 1786–93.

Lenton, Timothy M., Johan Rockström, Owen Gaffney, Stefan Rahmstorf, Katherine Richardson, Will Steffen, and Hans Joachim Schellnhuber. "Climate Tipping Points—Too Risky to Bet Against." *Nature* 27 (November 2019). https://www.nature.com/articles/d41586-019-03595-0.

Lewalski, Barbara. *The Life of John Milton: A Critical Biography*. Rev. ed. Oxford: Blackwell, 2002.

———. *Milton's Brief Epic: The Genre, Meaning, and Art of Paradise Regained*. Providence, RI: Brown University Press, 1966.

Leonard, John. *Faithful Labourers: A Reception History of Paradise Lost, 1667–1970*. 2 vols. Oxford: Oxford University Press, 2013.

Leonard, John, ed. *Paradise Lost*. London: Penguin, 2000.

Lewis, Elizabeth Jayne. *Air's Appearance: Literary Atmosphere in British Fiction, 1660–1794*. Chicago: The University of Chicago, 2012.

Lombard, Peter. *The Sentences, Book 2: On Creation*. Translated by Giulio Silano. Toronto: Pontifical Institute of Mediaeval Studies, 2008.

Long, A. A. *Hellenistic Philosophy: Stoics, Epicureans, Sceptics*. 2nd ed. Berkeley: University of California Press, 1986.

———. "Stoic psychology." In *Cambridge History of Hellenistic Philosophy*, edited by Algra, Barnes, Mansfeld, and Schofield, 560–84.

Longfellow, Henry Wadsworth. *The Golden Legend*. London, 1854.

Lucan, *The Civil War*. Translated by J. D. Duff. Cambridge, MA: Harvard University Press, 1928.

Lucretius. *De Rerum Natura*. Translated by W. H. D. Rouse. 2nd ed. Cambridge: Cambridge University Press, 1982.

———. *On the Nature of Things*. Translated by Cyril Bailey. Oxford: Clarendon Press, 1921.

Luther, Martin. *A Critical and Devotional Commentary on Genesis*. Vol. 1 of *The Precious and Sacred Writings of Martin Luther,* edited by John Nicholas Lenker. Minneapolis, MN: Lutherans in All Lands Co., 1904.

———. *Table Talk.* Edited and translated by Theodore G. Tappert. Vol. 54 of *Luther's Works,* edited by Helmut T. Lehmann. 55 vols. Philadelphia: Fortress Press, 1967.

Lynch, Rev. J. S. M. *Rite of the Blessing of a Bell, or of Several Bells, According to the Roman Pontifical.* New York: The Cathedral Library Association, 1912.

Lyons, Sir Henry. *The Royal Society, 1660–1940, A History of Its Administration under Its Charters.* Cambridge: Cambridge University Press, 1944.

Magnus, Albertus. *De Anima.* Edited by Clemens Stroick. Vol. 7 of *Opera Omnia.* Münster, Germany: Monasterii Westfalorum in Aedibus Aschendorff, 1955.

Mancosu, Paolo. "Acoustics and Optics." In *Cambridge History of Science,* vol. 3: *Early Modern Science,* edited by Katharine Park and Lorraine Daston, 596–631. Cambridge: Cambridge University Press, 2006.

Marcus, Leah. "Ecocriticism and Vitalism in *Paradise Lost.*" *Milton Quarterly* 49, no. 2 (2015): 96–111.

———. *The Politics of Mirth: Jonson, Herrick, Milton, Marvell, and the Defense of Old Holiday Pastimes.* Chicago: University of Chicago Press, 1986.

Marjara, Harinder. *Contemplation of Created Things: Science in* Paradise Lost. Toronto: University of Toronto Press, 1992.

Marshall, William. *The Rural Economy of Norfolk.* 2nd ed. 2 vols. London, 1795.

Martin, Craig. "Causation in Descartes' *Les Meteores* and Late Renaissance Aristotelian Meteorology." In *The Mechanization of Natural Philosophy,* edited by Roux and Garber, 217–36.

———. *Renaissance Meteorology: Pomponazzi to Descartes.* Baltimore: Johns Hopkins University Press, 2011.

Masson, David. *The Life of John Milton.* 6 vols. Cambridge: Macmillan and Co., 1871–94.

McClintock, John, and James Strong, eds. *Cyclopædia of Biblical, Theological, and Ecclesiastical Literature.* 10 vols. New York: Harper & Brothers, 1896.

McColl, R. W., ed. *Encyclopedia of World Geography.* Vol. 1 of 3. New York: Facts on File, 2014.

McColley, Diane Kelsey. *A Gust for Paradise: Milton's Eden and the Visual Arts.* Urbana: University of Illinois Press, 1993.

———. *Milton's Eve.* Urbana: University of Illinois Press, 1983.

McGuire, Maryann Cale. *Milton's Puritan Masque.* Athens: University of Georgia Press, 1983.

Mersenne, Marin. *Harmonie Universelle: The Books on Instruments.* Translated by Roger E. Chapman. The Hague, The Netherlands: Martinus Nijhoff, 1957.

———. *Harmonie universelle contenant la théorie et la pratique de la musique* [. . .]. [Paris], 1636.

Meyer, Susan Sauvé. "Chain of Causes: What Is Stoic Fate?" In *God and Cosmos in Stoicism,* edited by Ricardo Salles, 71–89. Oxford: Oxford University Press, 2009.

Meyer-Baer, Kathi. *Music of the Spheres and the Dance of Death: Studies in Musical Iconology.* Princeton: Princeton University Press, 1970.

Middleton, W. E. Knowles. *The History of the Barometer.* Baltimore: Johns Hopkins University Press, 1964.

Miller, Philip. *The Gardeners Dictionary: Containing the Best and Newest Methods of Cultivating and Improving* [. . .], vol. 2. 8th ed. London, 1768.
Minear, Erin. *Reverberating Song in Shakespeare and Milton: Language, Memory, and Musical Representation.* Surrey, England: Ashgate, 2011.
Nef, John U. *The Rise of the British Coal Industry.* 2 vols. London: George Routledge & Sons, 1932.
Neiman, Susan. *Evil in Modern Thought: An Alternative History of Philosophy.* Princeton, NJ: Princeton University Press, 2015. First printed in 2002.
Newton, Diana. *North-East England, 1569–1625: Governance, Culture and Identity.* Woodbridge, UK: Boydell & Brewer, 2006.
Nicholson, Majorie. "Milton and the Telescope." *English Literary History* 2.1 (April 1935): 1–32.
Nixon, Rob. *Slow Violence and the Environmentalism of the Poor.* Cambridge, MA: Harvard University Press, 2011.
Norbrook, David. "The Reformation of the Masque." In *The Court Masque*, edited by David Lindley, 94–110. Manchester: Manchester University Press, 1984.
Northcott, Michael S. *A Political Theology of Climate Change.* Cambridge: William B. Eerdmans Publishing Co., 2013.
[Oldenburg, Henry, and John Beale.] "Observations continued upon the Barometer, or Balance of the Air." *Philosophical Transactions* 1, no. 10 (March 12, 1665–66): 163–66.
Orgel, Stephen. "The Case for Comus." *Representations* 81, no. 1 (Winter 2003): 31–45.
Ortiz, Joseph M. *Shakespeare and the Politics of Music.* Ithaca, NY: Cornell University Press, 2011.
Oster, Emily. "Witchcraft, Weather and Economic Growth in Renaissance Europe." *Journal of Economic Perspectives* 18, no. 1 (Winter 2004): 215–28.
Parker, Geoffrey. *Global Crisis: War, Climate Change, and Catastrophe in the Seventeenth Century.* New Haven, CT: Yale University Press, 2013.
Partner, Jane. *Poetry and Vision in Early Modern England.* Early Modern Literature in History. Cham, Switzerland: Springer, 2018.
Pecquet, Jean. *New anatomical experiments of John Pecquet of Deip* [. . .]. Early English Books Online. British Library (Thomason, 194:E.1521[1]). London, 1653.
[Philaretes and John Hind.] *Work for Chimney-sweepers: or a warning for Tobacconists.* London, 1602.
Picciotto, Joanna. *Labors of Innocence in Early Modern England.* Cambridge, MA: Harvard University Press, 2010.
Planer, John H. "Damned Music: The Symbolism of the Bagpipes in the Art of Hieronymus Bosch and His Followers." In *Music from the Middles Ages through the Twentieth Century: Essays in Honor of Gwynn McPeek*, edited by Carmelo P. Comberiati and Matthew C. Steel, 335–56. New York: Gordon and Breach Science Publishers, 1998.
Plat, Hugh. *A new, cheape and delicate fire of cole-balles wherein seacole is by the mixture of other combustible bodies, both sweetened and multiplied* [. . .]. London, 1603.
Plato, *Republic*, vol. 9: *Books 6–10*. Edited and translated by Christopher Emlyn-Jones, William Preddy. Cambridge, MA: Harvard University Press, 2013.
———. *Timaeus.* In *Timaeus. Critias. Cleitophon. Menexenus. Epistles.*, translated by R. G.

Bury, 17–253. Vol. 7 of *Plato with an English Translation*. Cambridge, MA: Harvard University Press, 1952.

Pliny the Elder. *Natural History*, vol. 1: *Books 1–2*. Translated by H. Rackham. Cambridge, MA: Harvard University Press, 1938.

———. *Natural History*, vol. 9: *Books 33–35*. Translated by H. Rackham. Cambridge, MA: Harvard University Press, 1952.

Polsue, Joseph. *A Complete Parochial History of the County of Cornwall*. Vol. 2. Truro, UK, 1868.

Pontificale Romanum Clementis VIII. Paris, 1615.

Porta, Giambattista della. *Natural magick*. London, 1669.

Praetorius, Michael. *Syntagma Musicum II, De Organographia: Parts I and II*. Translated and edited by David Z. Crookes. Oxford: Clarendon Press, 1986.

Ptolemy. *Ptolemy's Tetrabiblos: Or, Quadripartite; Being Four Books of the Influence of the Stars*. Translated and edited by J. M. Ashmand. London: W. Foulsham & Co., 1917–36.

Quint, David. "The Disenchanted World of *Paradise Regained*." *Huntington Library Quarterly* 76, no. 1 (Spring 2013): 181–94.

"A Relation of some Mercurial *Observations and their Results*," *Philosophical Transactions* 1, no. 9 (February 12, 1665–66): 153–59.

Rogers, John. *The Matter of Revolution: Science, Poetry, and Politics in the Age of Milton*. Ithaca: Cornell University Press, 1996.

Rossi, Paolo. *Francis Bacon: From Magic to Science*. Translated by Sacha Rabinovitch. London: Routledge & Kegan Paul, 1968.

Roux, Sophie, and Daniel Garber, eds. *The Mechanization of Natural Philosophy*. Dordrecht: Springer, 2013.

Rowlands, Samuel. *Greenes ghost haunting conie-catchers*. London, 1602.

Rumrich, John, and Stephen M. Fallon. "Introduction." In *Immortality and the Body in the Age of Milton*, edited by Rumrich and Fallon, 1–20. Cambridge, UK: Cambridge University Press, 2018.

Rumrich, John. *Matter of Glory: A New Preface to* Paradise Lost. Pittsburgh: University of Pittsburgh Press, 1987.

Sambursky, Sam. *The Physical World of Late Antiquity*. Princeton: Princeton University Press, 1962.

A Sancto Paulo, Eustachius. *Tertia Pars Summa Philosophiae, Quae est Physica*. Vol. 2 of *Summa Philosophiae Quadripartite de Rebus Dialecticis, Moralibus, Physicis et Metaphysicis*. 2 vols. Paris, 1609.

Schott, Gaspar. *Magia Universalis Naturæ et Artis: sive, Recondita naturalium & artificialium rerum scientia* [. . .] *opus quadripartitum* [. . .]. 4 vols. Herbipoli [Würzburg], 1657–59.

———. *Mechanica Hydraulico-Pneumatica: qua Praeterquàm quòd Aquei Elementi natura, proprietas, vis motrix, atque occultus cum aëre conflictus, à primis fundamentis demonstratur* [. . .]. Fracofurtens. [Frankfurt am Main]: Sumptu heredum Joannis Godfridi Schönwetteri [. . .] Excudebat Henricus Pigrin Typographus, Herbipoli, 1657.

Seneca. *Natural Questions*, vol. 2: *Books 4–7*. Translated by Thomas H. Corcoran. Cambridge, MA: Harvard University Press, 1972.

Shakespeare, William. *The Comical History of the Merchant of Venice, or Otherwise Called the Jew of Venice.* In *The Norton Shakespeare*, edited by Stephen Greenblatt, 1090–145. New York: W. W. Norton, 1997.

———. *Macbeth*. Edited by Nicholas Brook. The Oxford Shakespeare, edited by Stanley Wells. Oxford: Oxford University Press, 2008.

———. *The Tempest*. Edited by Peter Holland. The Pelican Shakespeare, edited by Stephen Orgel and A. R. Braunmuller. New York: Penguin, 1999.

Shank, J. B. "What Exactly Was Torricelli's 'Barometer'?" In *Science in the Age of Baroque*, edited by Ofer Gal and Raz Chen-Morris, 161–96. International Archives for the History of Ideas 208. Dordrecht: Springer, 2018.

Shapin, Steven, and Simon Schaffer. *Leviathan and the Air-Pump: Hobbes, Boyle and the Experimental Life*. Princeton: Princeton University Press, 1985.

Sherry, Beverley. "Milton, Materialism, and the Sound of *Paradise Lost*." *Essays in Criticism* 60, no. 3 (2010): 220–41.

Shumaker, Wayne. *The Occult Sciences in the Renaissance: A Study in Intellectual Patterns*. Berkeley: University of California Press, 1972.

Simons, Patricia. *The Sex of Men in Premodern Europe: A Cultural History*. Cambridge: Cambridge University Press, 2011.

Sloterdijk, Peter. *Terror from the Air*. Translated by Amy Patton. Los Angeles: Semiotext[e], 2009.

Smith, Bruce. *The Acoustic World of Early Modern England: Attending to the O-Factor*. Chicago: Chicago University Press, 1999.

Smith, Pamela H. "Science on the Move: Recent Trends in the History of Early Modern Science." *Renaissance Quarterly* 62, no. 2 (Summer 2009): 345–75.

Smith, William, and Samuel Cheetham, eds. *Dictionary of Christian Antiquities*. Vol. 1 of 2. London, 1876.

Sohm, Philip L. "Palma Vecchio's *Sea Storm*: A Political Allegory." *RACAR: Revue d'Art Canadienne / Canadian Art Review* 6, no. 2 (1979–80): 85–96.

Spaeth, Sigmund. *Milton's Knowledge of Music*. Ann Arbor: University of Michigan, 1963. First published by Princeton University Library in 1913.

Starnes, DeWitt T., and Ernest William Talbert. *Classical Myth and Legend in Renaissance Dictionaries: A Study Of Renaissance Dictionaries in Their Relation to the Classical Learning of Contemporary English Writers*. Chapel Hill: University of North Carolina Press, 1955.

Steadman, John M. "Heroic Virtue and the Divine Image in *Paradise Lost*." *Journal of the Warburg and Courtauld Institutes* 22, no. 1/2 (1959): 88–105.

———. "'Like Turbulencies': The Tempest of 'Paradise Regain'd' as Adversity Symbol." *Modern Philology* 59, no. 2 (1961): 81–88.

Steggle, Matthew. "*Paradise Lost* and the Acoustics of Hell." *Early Modern Literary Studies* 7, no. 1 (2001). https://extra.shu.ac.uk/emls/07-1/stegmil2.htm.

Stjerna, Kirsi I., and Else Marie Wiberg Pedersen. "Introduction to Lectures on Genesis 1:26–2:3 and Genesis 2:21–25, 1535." In *The Annotated Luther*, vol. 6: *The Interpretation of Scripture*, edited by Euan K. Cameron, 67–77. Minneapolis, MN: Fortress Press, 2017.

St. Jerome. "Ephesians 6:12." In *The Commentaries of Origen and Jerome on St Paul's Epistle* to the Ephesians, edited by Ronald E. Heine, 254–60. Oxford: Oxford University Press, 2002.

Svendsen, Kester. "Milton's 'Aerie Microscope.'" *Modern Language Notes* 68, no. 8 (1949): 525–27.

Swaim, Kathleen M. "Hercules, Antaeus, and Prometheus: A Study of the Climactic Epic Similes in *Paradise Regained*." *Studies in English Literature* 18, no. 1 (Winter 1978): 137–53.

Tacitus, *The Histories: Books I-III*. Translated by Clifford H. Moore. The Loeb Classical Library. Cambridge, MA.: Harvard University Press, 1925.

Taub, Liba. *Ancient Meteorology*. London: Routledge, 2003.

Taylor, Jeremy. "Sermon XVII. The Marriage Ring [. . .]." In *A Course of Sermons for All the Sundays in the Year*, 207–19, vol. 4 of *The Whole Works of Right Rev. Jeremy Taylor*, edited by Rev. Charles Page Eden. 10 vols. London: Longman, Brown, Green, and Longmans, 1850.

Taylor, Jesse Oak. *The Sky of our Manufacture: The London Fog in British Fiction from Dickens to Woolf*. Charlottesville: University of Virginia Press, 2016.

Thommen, Lukas. *An Environmental History of Ancient Greece and Rome*. Translated by Philip Hill. Cambridge: Cambridge University Press, 2012.

Tigner, Amy. *Literature and the Renaissance Garden from Elizabeth I to Charles II: England's Eden*. Burlington, VT: Ashgate, 2012.

Todd, Henry J., ed. *The Poetical Works of John Milton with Notes of Various Authors*. 2nd ed. 7 vols. London: Law and Gilbert, 1809.

Treip, Mindele Anne. *Allegorical Poetics and the Epic: The Renaissance Tradition to Paradise Lost*. Lexington: University Press of Kentucky, 1993.

Trenberth, Kevin E. and Christian J. Guillemot. "The Total Mass of the Atmosphere." *Journal of Geophysical Research: Atmospheres* 99, no. D11 (November 20, 1994): 23079–88.

Trubowitz, Rachel. "'Nor vacuous the space': Milton's Chaos and the Vacuist–Plenist Controversy." In *A New Companion to Milton*, edited by Thomas M. Corns, 460–74. Chichester, UK: John Wiley & Sons, 2016.

Turner, Raymond. "English Coal Industry in the Seventeenth and Eighteenth Centuries." *The American Historical Review* 27, no. 1 (October 1921): 1–23.

United Nations. "United Nations Framework Convention on Climate Change (UNFCCC)." (1992): 1–33.

United Nations Environment Programme. *Emissions Gap Report 2019*, UNEP, Nairobi. http://www.unenvironment.org/emissionsgap.

Van De Pitte, Frederick P. "Some of Descartes' Debts to Eustachius a Sancto Paulo." *Monist* 71, no. 4 (October 1988): 487–97.

Verity, A.W., ed. "Appendix D: Paradise Lost, I.515–17." In *Paradise Lost*, 674–76. Cambridge: Cambridge University Press, 1910.

Walker, D. P. "The Astral Body in Renaissance Medicine." *Journal of the Warburg and Courtauld Institutes* 21 (1958): 119–33.

———. "Francis Bacon and *Spiritus*." In *Music, Spirit and Language in the Renaissance*, edited by Penelope Gouk, 121–30. London: Varorium Reprints, 1985.

———. *Spiritual and Demonic Magic: From Ficino to Campanella*. London: The Warburg Institute, University of London, 1958.

Wallis, John. "Account of some Passages of his own Life [1696–97]." In T. Hearne, *Peter Langtoft's Chronicle: As Illustrated and Improv'd by Robert of Brunne, from the Death of Cadwalader to the End of K. Edward the First's Reign*, cxl–clxx. Vol. 1 of 2. Oxford: 1725.

Walters, H. B. *Church Bells of England*. London: Henry Frowde, 1912.

Warde, Paul, and Tom Williamson. "Fuel supply and agriculture in post-medieval England." *The Agricultural History Review* 62, no. 1 (2014): 61–82.

Weart, Spenser R. *The Discovery of Global Warming*. Rev. ed. New Histories of Science, Technology, and Medicine, edited by Margaret C. Jacob and Jacob R. Weart. Harvard, MA: Harvard University Press, 2008.

Weber, Harold M. *Paper Bullets: Print and Kingship under Charles II*. Lexington: University of Kentucky Press, 1996.

Weber, Max. "Science as a Vocation." In *The Vocation Lectures*, edited by David Owen and Tracy B. Strong, and translated by Rodney Livingstone, 1–31. Indianapolis, IN: Hackett Publishing, 2004.

Webster, C. "The Discovery of Boyle's Law, and the Concept of the Elasticity of Air in the Seventeenth Century." *Archive for History of Exact Sciences* 2, no. 6 (December 31, 1965): 441–502.

White, Andrew D. *A History of the Warfare of Science with Theology in Christendom*. 2 Vols. New York: D. Appleton, 1915.

Wilkins, John. *The Discovery of a World in the Moone. Or, a Discourse tending to prove that 'tis probable there may be another habitable World in that Planet*. London, 1638.

———. *Mathematical Magick: or, The Wonders that May be Performed by Mechanichal Geometry. In Two Books*. 2nd ed. London, 1680.

Wilson, George. "On the Early History of the Air-Pump in England." *Edinburgh New Philosophical Journal*. Edinburgh: Neill and Company, 1849.

Winternitz, Emanuel. *Musical Instruments and Their Symbolism in Western Art*. New Haven: Yale University Press, 1979.

Wrigley, E. A. *Energy and the English Industrial Revolution*. Cambridge: Cambridge University Press, 2010.

Index

Accademia del Cimento, 113–14, 128
acoustics, 4–6, 10, 51, 195n39, 202n18; and atmosphere, 187–90; and audible species, 53–54, 61, 195n36; and breath, 42–43; and corruption, 46–47, 96–97, 171–72, 180–81; divine, 70–71; and experimentation, 51–52, 54–55, 72–73; and instruments, 109, 113–14; and metals, 99–101; and meteorology, 33–34, 38–39, 96, 100, 107–8, 111–18, 175; and natural magic, 49–55; and nature, 46–47, 109–10; and politics, 95; and Satan, 97–98, 101–2, 104, 106–8, 111–12, 117–18; and space, 29–30, 44–45. *See also* instruments; music; sound
action. *See* agency
Adams, Thomas, *The Devil's Banket*, 165–67
Addison, Joseph, 137
agency, 2–3; angelic, 85–86; demonic, 80, 83–87, 157–58, 175, 193n8; heroic, 76–77
Agrippa, Heinrich Cornelius, 47–51, 56–57, 62–63; *Three Books of Occult Philosophy*, 8–9
air, 5–8, 79; and body, 7–8; and breath, 1–2; and corruption, 10–11, 24–25, 42–45, 61, 78–79, 180–84; and demons, 90–93, 120, 154, 155–56; elasticity of, 133–38; as element, 6–8; and exhalations, 31–32, 35–37; and instruments, 108–15; materiality of, 5–7, 9–10, 125–28, 187–88; and mediation, 1–2, 7, 24–25, 33–34, 57–63; and meteorology, 79–80, 96; middle (region of), 35–36, 83–84, 90, 95–96, 176–77, 190–91, 206nn8–9; and music, 47; and nature, 35–36; and pollution, 1–2, 146–50, 219n8, 220nn9–10, 220n15; pressure of, 121, 132, 135–37, 140–43, 183, 187–88, 190–91, 214n7, 215n17; regions of, 80–81; and Satan, 3–4, 10–11, 79–85, 92–101, 104, 107, 109–17, 138–39, 151–59, 172–75, 177–78, 180–81; and science, 26–27; and soul, 5–6, 8–9; and sound, 1–2, 7–10, 24–25, 27, 29–30, 33–34, 44–45, 80, 91–93, 103–4, 116–17; and speech, 24–25, 30–31; weight of, 120–42, 190–91. *See also* atmosphere; breath and breathing; demons and demonology; exhalations; meteorology
alchemy, 107, 209n20
allegory, 13, 81–82, 86–87
Anaximander, 31–32
Anaximenes, 31–32
angels: body of, 88–92; and meteorology, 85–92. *See also* demons and demonology
animism, 3, 145; and breath, 42; and materialism, 193n11; vitalist, 79. *See also* matter and materialism; vitalism
apocalypse, 35–36, 150
Aquinas, Thomas, 84–86
Aratus, *Phaenomena*, 27–28
Archimedes, 139–40
Aristotle, 9–10, 47–49, 89–90, 122, 198n19, 198nn24–26, 199n34; *De Caelo* (On the Heavens), 123–25; *Meteorologica*, 26–28, 31–33, 197nn12–13; *Nichomachean Ethics*, 155–56, 171–72; theory of exhalations, 91–92, 99–100
Arnold, Clinton E., 83, 154

artifice, 101–2
astronomy, 47–48, 119–20, 127; and fluid-heaven theory, 127–28
Athanasius, 84–85; *The Life of Antony*, 112
atmosphere, 5–7, 79–80, 187; and acoustics, 8–9, 24–30, 39–40, 62–63, 69–70, 91, 96, 187–90; and corruption, 1–5, 12–13, 41–44, 76–77, 145, 154, 158–59, 162–63, 170–71, 177, 183–84; and demons, 154–57, 172–73; and instruments, 99–100; and morality, 78–79; and nature, 30–31, 35–36; as an ocean, 126–28, 216n52; and perception, 58; physics of, 120–21, 181–83, 190–91; and poetry, 189–90; regions of, 35–38, 41–42, 80–84, 90–91; and science, 26–27, 187–88; and soul, 24; and supernatural, 83–84; weight of, 190–91. *See also* air; pneumatics
atomism, 6–7, 43, 122, 188
Aubrey, John, 98
Augustine, St., 90; *The City of God against the Pagans*, 83–84, 155–56, 177

Bacon, Francis, 10, 47–51, 61–63, 66, 71–72, 111–12, 203n38; *New Atlantis*, 52–52; *Of the Dignity and Advancement of Learning*, 67–69; on spiritus, 52–53; *Sylva Sylvarum*, 51–52, 59–60, 67–69, 72–73, 113–14
Baliani, Giovanni Batista, 121–22
Barberini, Cardinal Francesco, 98
barometer, 5–6, 119–22, 181–84, 190–91, 213n6, 217n60; as scale, 138–43. *See also* air; pneumatics
Beale, John, 42–43, 127, 137–38, 183–84
Beeckman, Isaac, 124–25, 135–37
Berti, Gaspar, 122, 213n6
biblical commentary, 5–6; Acts, 55; Deuteronomy, 149, 175–76, 183; Ephesians, 79, 81–83, 95–96, 150–51, 153–54, 162–63, 173–74, 176–77; Exodus, 92–93, 160; Galatians, 157–58, 222n59; Genesis, 13–14, 221n27; Hebrews, 164; John, 158–59; Judges, 13–14; 1 Kings, 223n60; Luke, 158–59, 162; Matthew, 158–59; Numbers, 158–60; Psalms, 176–77

Bodin, Jean, *Universae Naturae Theatrum*, 84
body: and air, 7–8; and corruption, 180–81; demonic, 111–14; and instruments, 111–12; and organs, 98–99; and senses, 69–71; and soul, 23–24, 70–71, 74–76; and voice, 66–69
Boethius, 39
Bosch, Hieronymus, *Triptych of the Temptation of St. Anthony*, 84–85, 212nn62–63
Boyle, Robert, 5–6, 48–49, 125, 128–31, 133–35, 138–41, 182, 194n19, 218n75; Boyle's law, 119–20; *New Experiments Physico-Mechanical, touching the Spring and Weight of the Air*, 131–37
Brahe, Tycho, 47–48, 127
breath and breathing, 1–2, 7–8, 23–24, 39, 43, 193n4, 198n22; and acoustics, 42–45; and animism, 42–43; and contamination, 97–98; and instruments, 111; and meteorology, 31–32, 36–37, 42–43; and pollution, 146–47; satanic, 100–101; and soul, 31–32, 42–43. *See also* air
Bridgewater, Earl of, 46–47, 146–47, 220n16
brimstone, 147–50, 221n27. *See also* sulfur
Browne, Thomas, 147–48
Burton, Robert, 8, 10–11
Butler, Charles, *Principles of musik, in singing and setting*, 64–65

calamity, 2–3, 148–49
Callot, Jacques, 84–85; *The Temptation of Saint Anthony*, 112
cannons, 13, 96–97, 112–14
Catholicism, 84, 104–7, 144, 151, 219n2
Cavert, William, 161–62
Charleton, Walter, 135, 138–39; *Physiologica Epicuro-Gassendo-Charltoniana*, 135–37
chastity, 23, 46, 71–74, 175
Cicero, 22–23
climate, 1–3, 5, 7, 16–17, 79, 89, 196n55; and atmosphere, 5–7, 79–80; climate change, 1–3, 11–17, 189–91, 193n3, 196n46, 196n54, 196n57; climatic zones, 5, 194n17; corruption of, 187–88; and free will, 145; and morality, 1–4, 11–17,

78–79, 144–54, 183–84, 187–91; and original sin, 12–16; and prophecy, 15–17; and sound, 7–8, 91–94, 97–98, 107–8. *See also* ecocriticism; meteorology
clouds, 87–91
coal, 150, 154, 159–61; and food, 160–63; and pollution, 146–49, 162, 180–81, 219n8, 220nn9–10. *See also* fuel; pollution
Cockayne, Emily, 146
Comes, Natalis, *Mythologiae,* 180–81
condensation, 42–43, 92–93, 99–100. *See also* meteorology
consonance, 5–6, 9–10, 49, 108–9, 202n18. *See also* music; sound
Copernicus, Nicolaus, *De Revolutionibus Orbium Coelestium* (On the Revolutions of the Heavenly Spheres), 47–48
corpuscles, 6–7, 48–49, 141
Corpus Hermeticum, 157–58
corruption, 1–5, 10–11; atmospheric, 12–13, 24–25, 41–45, 145, 154, 158–59, 162–63, 170–71, 177, 180–81, 183–84; of climate, 187–88; environmental, 35–38, 46–47, 75–77, 157–59, 163–73, 179–80, 184–85; and exhalations, 91–92; and experimentation, 187–88; and sound, 41–45, 61–63, 74–77. *See also* demons and demonology; Satan
cosmography, 27–29, 119, 177–78

Dati, Carlo, *Lettera a Filaleti,* 128
Dee, John, 49–51
Democritus, 122. *See also* atomism
demons and demonology, 5–6, 10–11, 49–51; and air, 120, 154, 155–56; and elements, 157–59, 166–74, 177–81; and instruments, 101–2, 111–15; and metals, 99–101; and meteorology, 80–94, 100–101, 157–58, 175; and nature, 172–73; and pollution, 148–53, 162–63. *See also* body; corruption; Satan
demonstration. *See* experimentalism
Descartes, René, 47–48, 91, 124–25; *Principia Philosophaie* (Principles of Philosophy), 10, 48–49
Dickens, Charles, *Hard Times,* 151–53

Digges, Thomas, 47–48
Diodati, Charles, 28–29
dissonance, 46–47, 117–18, 174. *See also* music; sound
Drebbel, Cornelius, 103–4

ecocriticism, 189–90, 194n14
ecology: and crisis, 2–3, 147, 152–53, 189; and Fall, 3–5, 14–15, 145; and morality, 3, 159, 189–90
economy, 159–62, 219n8, 223nn71–72; and morality, 166; of spirit, 71
ecstasy, 23, 33–34, 69–71
Edwards, Karen L., 4, 79–80
elements, 6–8, 42–43, 126–28, 141–43; and harmony, 40; and meteorology, 89–91; stratification of, 36–37, 40, 120–21, 123–24, 199n36. *See also* air; *under* demons and demonology
Eliot, T. S., 5, 95
empire, 165–66, 169–73, 189–91
empiricism, 9–10, 50–51, 80, 103–4, 108–9
environment: corruption of, 157–59, 163–73, 179–80, 184–85; and demonic agency, 86–87, 157–58; and exhalations, 92; exploitation of, 3, 61–62, 85, 96, 156–57, 163, 173; and humans, 145; and morality, 3, 166–67; and original sin, 3–4; and perception, 46, 74; and sound, 8–9, 44–45; and spirit, 3
epic (genre), 1, 154–56
ethics: forbearance, 145; heroism, 155–56; of sacrifice, 190–91; temperance, 74–75. *See also* morality
Evans, J. Martin, 29
Evelyn, John, 179–81; *Fumifugium,* 146
exhalations, 91, 93–94, 176–77, 198n19, 198nn24–26, 198n28; and metals, 99–101, 209n20. *See also* air; breath and breathing; meteorology
experimentalism, 4, 10, 47–52, 54–55, 72–73, 119–20, 205n3; and corruption, 187–88. *See also* science
exploitation. *See under* environment

faith, 1–2, 144, 152–53, 155–56; and science, 2–4, 81–84. *See also* morality

Fall, 1, 3–5, 14–15, 145; and air, 120–21; and climate, 144–47, 179–80, 183–84, 187; and corruption, 74, 183–84, 187; and mechanics, 10; and meteorology, 3–4, 11–17; and sound, 95–96. *See also* original sin
Fallon, Stephen M., 145
Faunce, Abraham, 180–81
Ferdinand II, Grand Duke, 126
Ficino, Marsilio, 39, 47–48, 51–54, 62–63, 66–67, 70–71, 200n50; *De triplici vita* (Three Books on Life), 56–57
Fludd, Robert, 49–51
Forsyth, Neil, 153–54, 177
Frescobaldi, 98
Frye, Northrop, 177–78
fuel, 159–63, 193n2, 223n66; and crisis, 147, 159–61. *See also* coal; pollution

Galilei, Vincenzo, 49, 202n18
Galileo (Galileo Galilei), 5–6, 10, 119, 121–22, 202n18, 213n1; *Discorsi e dimostrazioni matematiche intorno à due nuove scienze* (Two New Sciences), 48–49, 121–22, 124–27, 213n6; *Siderius Nuncius* (Starry Messenger), 47–48
Garber, Daniel, 47–48
Gardiner, Ralph, 162–63
Gassendi, Pierre, 113–14
Gouge, William, *The vvhole-armor of God*, 150–53
Gouk, Penelope, 49–51, 107, 203n40, 205n65, 211n50; *Music, Science and Natural Magic*, 195n35, 195n37, 201n9, 202n14, 202n21, 202nn26–28, 203n31, 204n63
Great Fire (1666), 147, 150
Guerlac, Henry, 147–48
gunpowder, 13, 147–49, 151–54
Gunpowder Plot (1605), 151

Haak, Theodore, 128–31
hagiography, 84–85
Harding, David P., 135
harmony, 21–23, 29–30, 44–45, 47; and cosmos, 38–40; and *musica mundana*, 21–22, 195n38; and natural magic, 56–57; numerical, 51–52; and spiritus, 52–53; and vibrational behavior of strings, 49–51. *See also* music
Hartlib, Samuel, 128–31
hearing, 21–22, 51–54, 58–60, 70–72, 74, 76–77. *See also* acoustics; perception; sound
Henry, John, 49–51
Hequembourg, Stephen, 97
Hermetica (Pico della Mirandola and Ficino), 47–48
Hiltner, Ken, 3–4, 147
Hippocrates, 5
history and historiography, 1–3, 5, 80–82, 84, 171–72
Hoxby, Blair, 62–63
human and humanism, 5–6, 48–49, 144–45, 157–58, 164–66. *See also* nature
Hunt, Leigh, 101–2

idealism, 23
imagination, 23, 69, 95
instruments, 51–53, 66–69, 107, 156–57; activation by natural forces, 103–4; and air, 99–100, 108–12; bagpipes, 112–13, 212n65; church bells, 104–7; classical perceptions of, 101–2; and corruption, 183–84, 188; and demons, 101, 107–8, 111–15; horns, 101–2, 112, 115; organs, 5, 91, 93–94, 95, 97–118, 208n3, 209n11, 210n36, 213n82; and Satan, 98, 104–6, 109–11, 115–17; serpent, 115–16; and sound, 97–98, 104, 107–8, 111–12; virginal, 103–4; and voice, 115; wind, 6–7, 101–2, 108–12, 195n34. *See also* acoustics
Interregnum, 144–45

Jessey, Henry, *The exceeding riches of grace*, 149–50
Johnstone, Nathan, 10–11

Keckermann, Bartholomaeus, *Systema Systematum*, 27–28
Kepler, Johannes, 7–8, 47–48
Khunrath, Heinrich, *Amphitheatrum Sapientiae Aeternae*, 107
Kircher, Athanasius, 122; *Musurgia Universalis*, 103, 115

knowledge: divine, 4, 25–26; and magic, 55–56; of nature, 25–27, 47–48; theoretical, 48–49

Latour, Bruno, 2–3
Lawrence, Henry, *Of our communion and warre with angels*, 150–51
Leavis, F. R., 95
Licensing Act (1662), 144–45
life, 15–17, 40, 44, 78–79. *See also* air; breath and breathing
Little Ice Age, 1–2, 159–60, 193n3, 194n16. *See also* climate; fuel
Lombard, Peter, *Sentences*, 90
Longfellow, Henry Wadsworth, *The Golden Legend*, 106
Lucan, *The Civil War*, 178–79
Lucretius, 43, 66–67; *De rerum natura* (On the Nature of Things), 28
Luther, Martin, 11–12, 84–86

Macrobius, 22–23
magic, 3; ceremonial, 56–57, 72–73, 104–6, 202n21; and corruption, 62–63, 72–73; and experimentation, 72–73; and instruments, 51–52, 67–69; and knowledge, 55–56; and mechanics, 102–3; natural, 8–9, 49–55, 203n38; and science, 47–48, 61; sorcery, 55–56, 62, 67–69, 76–77; and sound, 49–51, 56–57, 61–69, 72–73, 103; and spiritus, 52–53
Magiotti, Rafael, 122
Magnus, Albertus, 101, 104
Marcus, Leah S., 3–4
Marjara, Harinder, 139, 142–43
masque (genre), 46–47, 62–63, 200n1; and music, 53–54
Masson, David, 22–23
mathematics, 47–51, 101–2, 139–40, 194n13; of music, 8–10, 39–40, 49–51, 53, 108–9
matter and materialism, 3, 10, 15–16, 32, 42, 52–53, 79, 145, 193n11, 205n3; and air, 187; and mechanics, 10, 96–97; and mixture, 89–91; and poetry, 1–2; and soul, 23, 145; and sound, 96–97. *See also* animism; monism; vitalism

mechanics and mechanical philosophy, 10, 101–2, 108–9, 139–40, 188; and atmosphere, 187–88; and magic, 102–3; and pneumatics, 103–4; and sound, 111–12
media and mediation: and air, 7, 16–17, 24–25, 33–34, 56–63; and music, 49–51, 53; and water, 72–73
Melanchthon, Philip, 84
Mersenne, Father Marin, 49–51, 101–2, 113–14, 126, 128–31, 202n18, 216n44; *Harmonie universelle*, 6–8, 108–11, 115
metals, 99–101, 104–6, 209n20
meteorology, 1–6, 89, 100, 194n13, 194n19, 197n13, 199n36, 205n3; and acoustics, 33–34, 38–39, 91–94, 96–98, 107–8, 111–18, 175; and angelic agency, 85–89; and atmosphere, 36–38, 79–80; and barometer, 121; clouds, 10–11, 32–37, 43, 58, 62–63, 80–82, 85, 87–94, 146, 151, 198n19, 220n20, 221n24; comets, 16–17, 79–80, 84, 91, 100, 183–84, 189–90, 198n19; and demonic agency, 80–94, 97–98, 100–101, 111–12, 120, 157–58, 175; dew, 25–26, 37, 42–44, 79–80, 87, 91–93, 206n8; earthquakes, 3–4, 16–17, 91, 147–48, 157–58, 198n25, 199n34, 207n44; and exhalations, 31–33, 36–37, 42–43, 91–94, 99–101, 147–48, 176–77, 179–80, 198nn24–26; fog, 62, 64–65; lightning, 10–11, 24, 81–82, 84, 87, 90–92, 150–51, 156, 174–75, 178–80, 183–84, 199n40; meteors, 10–11, 27–28, 84, 148–49, 183–84, 198n24; and morality, 2–3, 78–80, 84; and original sin, 2–4, 10–17; and pneuma, 198n22; snow, 24–26, 30–34, 37–38, 43–44, 90–91, 147–48, 198n19; and spirit, 79, 183–84; sulfur-nitre theory, 147–49; thunder, 24–26, 37, 43–44, 84, 87, 90–91, 105–6, 112–14, 147–51, 157–58, 178–80, 199n40, 208n47, 208n49; and uncertainty, 25–27; wind, 13, 15–16, 25–26, 31–38, 84–91, 93–94, 109–12, 163–64, 174–75. *See also* climate; storms and storm-raising
Michelangelo, *The Torment of Saint Anthony*, 84–85
microscopes, 156–57, 173–74, 183, 222n55. *See also* telescopes

250 INDEX

middle air. *See under* air
Middleton, W. E. Knowles, 122
Milton, John, 1–3; and acoustics, 5, 7–8, 10, 38–40, 43–44, 51, 53–54, 59–61, 66, 70–71, 95, 199n43, 208nn3–4; and allegory, 81–82, 144–45; and blindness, 95, 146–47; and climate, 5, 11–12, 80–81, 187–91, 194n14, 209n8; and contemporary perspectives, 4, 9–10, 48–49, 144–45, 187–91, 201n13, 205n3, 219n2; on corruption, 10–11; and exhalations, 91–94; and experimentation, 54–55; and flight, 132–35, 137–38; and Galileo, 119, 121, 213n1; and heroism, 155–56, 175–77; in Italy, 125, 128; in London, 146–47; and meteorology, 27–28, 79–80, 83, 87–88, 127–28, 128–31; monism of, 15–16, 145; on *musica mundana*, 21–23, 27; and nautical imagery, 110–11; and organ music, 98–99, 101–2; on original sin, 1–4, 11–16; and pneumatics, 132, 138–43, 214n7; youth, 21–22
Milton, John, works by: *Ad Patrem* (To my father), 21, 26–27; *Areopagitica*, 55, 213n1; "At a Solemn Music," 12, 47; *At a Vacation Exercise*, 21, 24, 29; *Comus*, 43–44, 46–49, 51, 53–77, 125, 175, 187–90, 200nn1–2, 201n13; *De Doctrina Christiana*, 14–15, 81–82, 193n4; "Elegia sexta [Elegy VI]," 28–29; *An Epitaph on the Marchioness of Winchester*, 197n2; *In Obitum Praesulis Eliensis* (On the Death of the Bishop of Ely), 197n2; *Lycidas*, 197n2; *Of Education*, 28, 98, 107, 128–31; *On The Morning of Christ's Nativity* (Nativity Ode), 12, 21, 27–47, 165, 172–73, 183–84, 187, 189; *Paradise Lost*, 1, 3–4, 7–8, 10–17, 37–38, 63–64, 70–71, 78–121, 123, 127–28, 132–39, 141–43, 147–49, 156–57, 180–81, 183–84, 187–90; *Paradise Regained*, 3–4, 10–11, 81–82, 91, 95–96, 110–11, 144–45, 153–91; *The Passion*, 21, 27; *Il Penseroso*, 42–43, 157–58; *Poems*, 128; *Prolusion II* ("On the Harmony of the Spheres"), 21–23, 27–28; *Prolusion VII* ("Learning brings more Blessings to Men than Ignorance"), 22–23, 25–27, 56; *A Readie & Easie Way to Establish a Free Commonwealth*, 219n4; *Samson Agonistes*, 144–45; "Sonnet XVI," 167; *The Second Defense of the English People*, 28, 887–88; *Upon the Circumcision*, 21
Minear, Erin, 96–97
monism, 10, 15–16, 145, 205n3. *See also* matter and materialism; vitalism
morality, 2–3, 193n8; and atmosphere, 78–79; and climate, 1–2, 13–14, 144–45, 147–54, 187–91; and environment, 2–3, 166–67; and fuel crisis, 159–61; and meteorology, 3–4, 11–12, 16–17, 78–80, 84; and sound, 10, 73–77, 97–98; and weight, 132
motion, 15–16, 66
music, 7–8, 52–54, 199n43, 200n50; and air, 7–10, 47; and automation, 103–4; and breath, 39; and consonance, 108–9, 202n18; divine, 27, 33–40, 64–66, 70–71; and ecstasy, 69–71; and harmony, 8–10, 21–22, 38–40, 47, 49, 195n38; and magic, 56–57, 67–69, 103; as medium, 49–51; *musica mundana*, 21–23, 195n38; and poetry, 95; and science, 49; and soul, 8–9, 39, 53–54, 107. *See also* acoustics; air; sound

nature, 2–5, 38–39, 43–45, 164–66; and acoustics, 109–10; and agency, 2–3; and air, 30–31, 35–36; corruption of, 37–38, 163–67, 172–73, 193n8; disenchantment of, 3, 28–29, 175; and exhalations, 35–37, 41–43; and harmony, 38–40; and human, 164–66; and knowledge, 25–26; and magic, 49–51; and musical instruments, 103–4; and original sin, 11–12; and speech, 33–35. *See also* environment; vitalism
Neoplatonism, 8–9, 22–23, 47–48, 69
New Science. *See under* science
Newton, Isaac, 147–48
Newton, Thomas, 137
Nicholson, Marjorie, 119
Nixon, Rob, 15
Northcott, Michael S., 2–3

occult, 47–56, 72–73, 80
Oldenburg, Henry, 182
organ (bodily), 72–73, 87–88, 93–94, 97–99, 112, 180–81, 195n36
organism, 1–2, 7, 15–16, 23–24, 32, 80, 104, 111, 145, 150–51
original sin, 1–5, 11–17, 91; and atmosphere, 24–25; and climate change, 12–17, 145, 174, 183–84, 187–91; and pollution, 147
outrage, 13–16, 37–38

Pandaemonium, 93–94, 100–110, 210n34
Pascal, Blaise, 128–31, 216n44
peccatogenesis, 11–12, 196n48
Pecquet, Jean, 125, 133–37
perception: aural, 52–54, 58–60, 63–64; and spirit, 69–72, 74–75; unreliability of, 46–47, 57–63, 76–77; and weather, 87–88
Perrier, Florin, 128–31
philosophy: Hermetic, 157–58; and magic, 47–48; mechanical, 47–48, 52–54; natural, 4–5, 9–10, 25–27, 48–49, 80, 84; of occult, 47–48, 52–53, 55–56, 72–73, 80; practical, 101–2
physics: of atmosphere, 119–20, 125–27, 134–37, 181–82; invention of barometer, 119–22; of weight, 123–43, 181–83, 190–91. *See also* science
Picciotto, Joanna, 4, 79–80
Pico della Mirandola, Giovanni, 47–48
Plato, 39, 42–43, 199n34; *Timaeus*, 53–54
Pliny the Elder, 178–79, 199n40; *Naturalis historia* (Natural History), 28, 36–37, 40
pneuma, 8, 23–24, 31–32, 39, 195n30, 197n9, 198n22. *See also* breath and breathing; music; spirit
pneumatics, 5–6, 8, 23–24, 31–32, 39, 71–72, 119–20, 122, 126, 131–32, 135, 137, 156–57, 213n6, 218n75; and atmosphere, 187; and mechanics, 103–4; and pressure, 134–43, 181–83, 190–91, 194n19; and spirituality, 183–84. *See also* air; science
poetry, 144–45; and atmosphere, 189–90; materiality of, 1–2; and science, 4; and sound, 5, 10, 95

politics, 193n3; and acoustics, 95, 101; of masque, 46–47; and nature, 2–3, 37–38; and poetry, 144–45
pollution, 1–2, 146–49, 162–63, 179–81, 219n8, 220nn9–10, 220n15; and demons, 148–53. *See also* coal; fuel
Pope, Alexander, *The Rape of the Lock*, 81–82
Porta, Giambattista della, 49–51; *Magia Naturalis*, 51–52
Pound, Ezra, 95
prolepsis, 16–17
prophecy, 5–6, 15, 46–47, 69, 132–33, 199n40, 200n50
Protestantism, 3, 10–11, 221n29
Ptolemy, 123–24, 127
Pythagoras, 8, 21–24, 39, 49

Quint, David, 172–73, 175

Ranelegh, Lady, 131–32
Renaissance, 3, 8–10, 28, 39, 47–51, 79–80, 84–85, 89, 100, 104, 115–16, 155–56, 189–90
resonance, 66, 101, 106, 204n63, 209n8. *See also* acoustics; music
Rete Medicea, 121
Ricci, Michelangelo, 126, 128–31
Roberval, Giles Persone de, 128–31
Rogers, John, 79–80
Royal Society, 48, 127, 131–32, 183–84
Rumrich, John, 132, 145

Satan, 88–89; and air, 79–85, 92–93, 95–101, 104, 107, 109–12, 114–17, 138–39, 151–59, 172–75, 177–78, 180–81; and environment, 158–59, 179–80; and exhalations, 176–77; instruments of, 98, 107, 109–10, 115–17; and luxury, 163–72; and meteorology, 86–87, 93–94, 97–98, 147–53, 158–59, 162–63, 172–75, 179–80, 194n16; and nautical, 85; passions of, 89–90; and sound, 95–98, 101–2, 107–12, 116–18, 171–72, 180–81; and stones, 176–77; voice of, 91–94, 97, 115–17, 180–81. *See also* corruption; demons and demonology

Scholasticism, 8–9, 47–48, 84, 89, 127, 142–43, 201n9; horror vacui, 122
Schott, Gaspar: *Magia Universalis Naturae et Artis*, 103; *Mechanica Hydraulico-Pneumatica*, 103
science, 3–4; and atmosphere, 5–6, 187–88; and corruption, 187–88; and magic, 47–51, 61; of meteorology, 187–88, 194n13, 194n19; mixed mathematical, 48–51; and nature, 3, 25–26, 28–29; New Science, 3–4, 48–49, 52–53, 121–26, 139–40, 183–84, 187–88, 190–91; practical, 48–49, 51; proto-sciences, 3, 194n12; of sound, 9–10, 49–51; and space, 26–27. *See also* acoustics; mathematics; meteorology; physics; pneumatics
Seneca, 32, 199n34; *Naturales quaestiones* (Natural Questions), 28
Sennert, Daniel, 147–48
Shakespeare, William: *Macbeth*, 7; *The Tempest*, 7–8
Sherry, Beverly, 95
Simplicius, 123–24
1645 Group, 128–32
Smith, Bruce, 117–18
smoke, 1–2, 91–92, 106, 112–13, 146–54, 179–81, 220nn9–10, 220n17. *See also* coal; fuel; pollution
soul. *See* spirit
sound, 1–2, 21–22, 29, 53, 57–60, 63–64, 95–97, 210n38; and air, 7–10, 24–25, 27, 29–30, 33–34, 53–54, 57–61, 69–70, 80, 91, 103–4, 116–17; and body, 21–22, 117–18; and corruption, 41–47, 61–63, 72–77, 91; and demons, 120; divine, 70–71; and environment, 44–47, 61–63, 72–77, 91; and experience, 54–55; and harmony, 21–23, 38–40; and instruments, 51–52, 97–98, 104, 107–8, 111–12; and magic, 56–57, 61–69, 72–73; materiality of, 96–97; and mechanical philosophy, 10, 49–51, 53–54, 109, 111–12; and metals, 101; and morality, 10, 73–77, 97–98; and motion, 66; pitched, 49–51; and poetry, 5, 95; satanic, 95–98, 101–2, 104, 107–12, 116–17, 171–72, 180–81; and science, 51; and sin, 174; and species, 61–63, 195n36, 203n40, 203n43; speed of, 59–61, 113–14; and spiritus, 52–54, 66–72; and voice, 56–57; and water, 72–73; and weather, 97–98; and writing, 117–18. *See also* acoustics; hearing; music; perception
space, 26–30, 174
species, 52–54, 60–63, 66, 174, 195n36, 203n33, 203n43, 204n57. *See also* acoustics; sound
speech and speaking, 24–25; and air, 30–31; and nature, 33–35. *See also* voice
Spenser, Edmund, *The Faerie Queene*, 137
spirit, 3, 31–32, 197nn6–7; and air, 8–9; and body, 23–24, 69–71, 74–76; and divine, 22–23; and environment, 3; materiality of, 23, 145; and music, 22–23, 39, 107; and sound, 70–76; and sympathetic resonance, 66; and voice, 66–67, 75–76. *See also* pneuma
Steggle, Matthew, 95
Stoicism, 8, 23–24, 39
stones, 176–77
storms and storm-raising, 1, 3–4, 10–15, 43, 84–85, 105–6, 109–11, 121, 147–51, 154–55, 159, 172–75, 179–81, 194n16, 199n40, 211n43, 225n103, 225n107; tempests, 7–8, 25, 78–79, 87, 105–6, 109–10, 150–51, 175. *See also* meteorology
sulfur, 14–15, 93–94, 146–49, 178–80, 209n20, 221nn24–25, 221n27. *See also* brimstone
suspension, 59, 62–63, 85, 132–34, 141–42
Swaim, Kathleen M., 177–78
sympathy, 52–53, 66. *See also* resonance

Taylor, Jesse Oak, 151–53
telescopes, 119–20
tempests. *See under* storms and storm-raising
Thales of Miletus, 84
Thyer, Robert, 137
tobacco, 151–54. *See also* smoke
Tomkins, John, 98
Torricelli, Evangelista, 121–22, 125–32, 134–41, 213n6, 216n44

Treip, Mindele Anne, 81–82
Trismegistus, Hermes, 42–43

vacuum, 121–22, 126–31, 134–35, 218n75
Vecchio, Palma, *Sea Storm*, 84–85
Virgil, 135, 167–68
virginity, 30–31, 64–66, 71–75
vitalism, 3, 8–9, 52–53, 79, 91–92, 96–97, 205n3. *See also* animism; matter and materialism
Viviani, Vincenzo, 126, 214n16
voice, 24–25, 29; and air, 30–31, 33–34, 44–45; and body, 66–69; and climate, 189–90; and corruption, 35–36; and harmony, 47; as instrument, 115–16; of nature, 38–39, 43–45; of Satan, 91, 97, 115–17, 180–81; and sound, 56–57; and spirit, 66–67, 75–76; and weather, 91–93. *See also* sound

Wallis, John, 131–32
water pumps, 122
weather. *See* climate; meteorology
Wilkins, John, *Mathematical Magick,* 101–4, 139–40, 210n38
will and free will, 3, 82, 84, 145. *See also* agency
Wren, Christopher, 131–32

Recent books in the series
UNDER THE SIGN OF NATURE: EXPLORATIONS IN ECOCRITICISM

Jeremy Chow • *The Queerness of Water: Troubled Ecologies in the Eighteenth Century*

Monica Seger • *Toxic Matters: Narrating Italy's Dioxin*

Taylor A. Eggan • *Unsettling Nature: Ecology, Phenomenology, and the Settler Colonial Imagination*

Samuel Amago • *Basura: Cultures of Waste in Contemporary Spain*

Marco Caracciolo • *Narrating the Mesh: Form and Story in the Anthropocene*

Tom Nurmi • *Magnificent Decay: Melville and Ecology*

Elizabeth Callaway • *Eden's Endemics: Narratives of Biodiversity on Earth and Beyond*

Alicia Carroll • *New Woman Ecologies: From Arts and Crafts to the Great War and Beyond*

Emily McGiffin • *Of Land, Bones, and Money: Toward a South African Ecopoetics*

Elizabeth Hope Chang • *Novel Cultivations: Plants in British Literature of the Global Nineteenth Century*

Christopher Abram • *Evergreen Ash: Ecology and Catastrophe in Old Norse Myth and Literature*

Serenella Iovino, Enrico Cesaretti, and Elena Past, editors • *Italy and the Environmental Humanities: Landscapes, Natures, Ecologies*

Julia E. Daniel • *Building Natures: Modern American Poetry, Landscape Architecture, and City Planning*

Lynn Keller • *Recomposing Ecopoetics: North American Poetry of the Self-Conscious Anthropocene*

Michael P. Branch and Clinton Mohs, editors • *"The Best Read Naturalist": Nature Writings of Ralph Waldo Emerson*

Jesse Oak Taylor • *The Sky of Our Manufacture: The London Fog in British Fiction from Dickens to Woolf*

Eric Gidal • *Ossianic Unconformities: Bardic Poetry in the Industrial Age*

Adam Trexler • *Anthropocene Fictions: The Novel in a Time of Climate Change*

Kate Rigby • *Dancing with Disaster: Environmental Histories, Narratives, and Ethics for Perilous Times*

Byron Caminero-Santangelo • *Different Shades of Green: African Literature, Environmental Justice, and Political Ecology*

Jennifer K. Ladino • *Reclaiming Nostalgia: Longing for Nature in American Literature*

Dan Brayton • *Shakespeare's Ocean: An Ecocritical Exploration*

Scott Hess • *William Wordsworth and the Ecology of Authorship: The Roots of Environmentalism in Nineteenth-Century Culture*

Axel Goodbody and Kate Rigby, editors • *Ecocritical Theory: New European Approaches*

Deborah Bird Rose • *Wild Dog Dreaming: Love and Extinction*

Paula Willoquet-Maricondi, editor • *Framing the World: Explorations in Ecocriticism and Film*

Bonnie Roos and Alex Hunt, editors • *Postcolonial Green: Environmental Politics and World Narratives*

Rinda West • *Out of the Shadow: Ecopsychology, Story, and Encounters with the Land*

Mary Ellen Bellanca • *Daybooks of Discovery: Nature Diaries in Britain, 1770–1870*

John Elder • *Pilgrimage to Vallombrosa: From Vermont to Italy in the Footsteps of George Perkins Marsh*

Alan Williamson • *Westernness: A Meditation*

Kate Rigby • *Topographies of the Sacred: The Poetics of Place in European Romanticism*

Mark Allister, editor • *Eco-Man: New Perspectives on Masculinity and Nature*

Heike Schaefer • *Mary Austin's Regionalism: Reflections on Gender, Genre, and Geography*

Scott Herring • *Lines on the Land: Writers, Art, and the National Parks*

Glen A. Love • *Practical Ecocriticism: Literature, Biology, and the Environment*

Ian Marshall • *Peak Experiences: Walking Meditations on Literature, Nature, and Need*

Robert Bernard Hass • *Going by Contraries: Robert Frost's Conflict with Science*

Michael A. Bryson • *Visions of the Land: Science, Literature, and the American Environment from the Era of Exploration to the Age of Ecology*

Ralph H. Lutts • *The Nature Fakers: Wildlife, Science, and Sentiment*

www.ingramcontent.com/pod-product-compliance
Lightning Source LLC
Chambersburg PA
CBHW021350300426
44114CB00012B/1163